10/17/11

SHOOTING STAR

The First Attempt By A Woman To Reach Hawaii By Air

By

Richard DuRose

Table of Contents

FOREWORD

In 1937, Amelia Earhart and her navigator were lost on a flight over the Pacific Ocean during her attempt to be the first woman to fly around the world. Her story, one of courage and tragedy, has become an enduring anecdote of American history.

Less known is the story of Mildred Doran a young Michigan school teacher. In 1927, she set her sights on being the first woman to fly from the West Coast of the United States to Hawaii, a distance of 2,400 miles. She was a participant in the Dole Transpacific Air Race which, promised fame and fortune to the first aviator to land at Honolulu from Oakland, California. James Dole, the "Pineapple King" initiated the event, giving the Dole Racers only eleven weeks from the time of his announcement until the start of the race. The racers had to acquire a plane, outfit it for the long journey, and get to the Oakland airfield for the start. During that eleven-week period, Mildred Doran and the other Dole Racers became celebrities, known and admired around the world. Mildred's sudden loss was tragic and regretted by all.

Mildred Doran was my mother's older sister, my aunt. In 1927, my mother was only twelve years old. When my mother died, I inherited pictures, newspaper articles, a family scrapbook, and other memorabilia from the Dole Race. Everything was stored away gathering dust. Then, one day I got them out for a closer look. I was immediately intrigued by the story. As a result, I present this account as homage to Mildred Doran. The Dole Race is an interesting chapter in the story of Flying's Golden Age. Here, I tell the whole story of that race, but with particular emphasis on the airplane named *Miss Doran* and its crew, Augie Pedlar, the pilot, Cy Knope, the navigator, and Mildred Doran.

1

AVIATION'S EARLY YEARS

It was near the end of 1903, in December, when the Wright brothers, from Dayton, Ohio, Orville and Wilbur, made their first successful flight in a powered aircraft on the beach at Kitty Hawk, North Carolina. Their first flight lasted less than one minute. It was a momentous event largely ignored. The only newspaper to cover the feat was the local North Carolina paper. But as we now know, it was the birth of aviation.

About a decade later, World War I began in Europe. The Allies and the Germans realized that airplanes could be used effectively in war. Initially, they were used to scout the movements of troops on the ground. Later, the planes were armed with guns and bombs. All during the War, improvements were made to airplanes which allowed for faster and longer flights. Americans became familiar with the brave American aviators of the War, such as Billy Mitchell, the father of the US Army Air Force, and Eddie Rickenbacker who shot down twenty two enemy planes. Even the enemy aviators became famous, such as the German pilot Manfred von Richthoven (later known as the Red Baron) with eighty kills.

World War I took the lives of many young men. Troops were thrown into protracted front line campaigns. Battles lasted for months, and lives were lost on both sides. 'Live now – for tomorrow you may die' was the adopted philosophy of many young adults of the time. In addition to the men who died in battle, there were many more sent home with serious and permanent disabilities caused by gunfire and shelling, as well as from mustard gas attacks.

Young women, patriotically assisting the war effort, went to work performing jobs previously held only by men. With the prospects for marriage lessened by a shortage of young men, women became much more independent. Before the War, women had few choices other than to be stay at home wives. They sought to keep their jobs for economic independence since marriage was no longer a sure thing. Women attended colleges in greater numbers. The suffrage movement triumphed and brought the 19th amendment to the Constitution giving women the right to vote in 1920. The "Gibson Girl" from before the war, with long flowing hair and long straight skirts, was replaced by the "Flapper," with short hair and short skirts. Coco Chanel introduced the androgynous look of short hair, low waist, and casually elegant comfortable clothes. New dances such as the Charleston and the Lindy Hop, required free movement, so Flappers did not wear corsets. Women became more assertive socially and, for the first time, openly smoked cigarettes in public. Late night parties were not to be missed. The popular novel, *Lady Chatterly's Lover* celebrated a new attitude toward sexuality in women. Prior to the '20s, women were supposed to remain in the background of life. During the '20s women sought to break free of old fashioned, stifling restrictions. This is not to say that women were free from stereotyped ideas as to their "place." But, in comparison with pre-war years, women were beginning to publicly aspire for equality.

At the end of World War I, thousands of airmen released from the military fanned out across the United States to earn a living showing off their skills and daring. Pilots trained during the War, snapped up thousands of surplus Curtis JN-4's, nicknamed the "Jenny" and fanned out across the country. Flying in the early 1920s was largely stunt flying. Fliers dazzled audiences with their exploits. A favorite stunt was to dive the plane straight toward the ground, only to pull up at the last second. Then, pilots added flying loops and wing walking to their exhibitions. In Paris, fliers flew through the *Arc de Triomphe* in front of an awestruck audience. Hundreds of WWI veterans introduced flying to the public with their barnstorming shows which toured small towns across the U.S. like carnivals. Indeed, the pilots were regularly referred to as "flying gypsies," or "barnstormers." People could read of the exploits of serious aviators and explorers such as Commander Byrd, but they could experience the thrill of flying personally at the flying gypsy shows.

Many of the reckless barnstormers died, literally at their own hand. To pull in larger audiences, flyers performed many outrageous stunts, such as wing walking, wing walking while the plane was in a dive, and even walking from the wing of one plane to another as the two planes sped only a few feet above the audience below. To many, flying appeared to be a reckless endeavor.

As time progressed, the image of fliers improved. A new breed of pilot took safety more seriously. Flying was used by explorers to visit hostile environments. Every new speed and distance feat of fliers was given first page newspaper headlines across the United States. Flying was beginning to capture the imagination of the country. This period is referred to as Flying's Golden Age.

During the second half of the '20s, airplane manufacturers sprang up all around the country. There were manufacturers in Hammondsport, New York; Wichita, Kansas;

Lincoln, Nebraska; Troy, Ohio; Los Angeles, California; Detroit, Michigan; and elsewhere. These companies had difficulty in filling their many orders for new planes. It was a rich man's sport to fly. But, there were plenty of rich men.

The 1920s were dubbed the "Roaring Twenties" for a reason. The economy was strong. Unemployment hovered around three per cent. Businesses were thriving. The stock market was on a steep upward swing. Many borrowed all they could to invest in stocks. Millionaires were created on a daily basis. The stock market was frothy, as we know now, and headed for a crash in 1929. But up until then, it was a period of unbounded optimism.

Newspapers were at, or near their most prolific in the '20s. They were the primary source of news. To increase circulation, newspapers tended to sensationalize their stories. Newspapers were sold on the streets. It was not unusual for "Extras" to be published to report on so-called important developments. Among my mother's memorabilia, is a scrapbook with 133 pages of newspaper clippings about Mildred and the Dole Race.

In the '20s, airplanes began to be used to transport the mail. They also were used on a limited basis for transportation. Early planes were typically two winged planes; biplanes constructed of canvas over a wood frame. By the end of the '20s, the newer models of airplane had a shell of metal and were single winged monoplanes.

Aviators made headlines on a regular basis. In 1922, Jimmy Doolittle made the first transcontinental flight from Pablo Beach (now known as Jacksonville Beach), Florida, to San Diego, California, with only one refueling stop. In 1926, Cdr. Richard Byrd, the famous explorer and his pilot, Floyd Benett, flew over the North Pole, although it was later questioned whether he actually made it all the way. In 1927, Byrd and a crew of three flew across the Atlantic from Roosevelt Field in Long Island to Normandy. On March 13, 1927, Theodore Gildred took off from San Diego and made the 4,200-mile flight with thirteen stops to Quito, Ecuador.

Then in May 1927, Charles Lindbergh flew his Ryan monoplane, *The Spirit of St. Louis*, non-stop for thirty-three and a half hours from New York's Long Island to Paris. He was not the first to fly non-stop to Paris, but the first to fly solo. *Lucky Lindy* became the world's most celebrated hero, and *The Spirit of St. Louis* the world's most famous airplane. Huge parades in his honor were staged in Paris and New York. In France, he received the "Medal of Honor." In Washington, he was given the "Distinguished Flying Cross." *The Lone Eagle's* flight set off a chain reaction. Everyone, it seemed, wanted to send Air Mail letters, travel by air, or learn to fly. Like space exploration and the astronauts in the 1960s, airplanes and their pilots inspired the world's imagination in the '20s. To this day, Lindbergh's triumph is recognized as one of the most important flights in aviation history.

As it happened, Henry Ford caught the flying bug. Ford had made a fortune selling his Model T automobiles, the first auto to be mass produced and reasonably priced. Ford took his first plane ride with Lindbergh at the controls. Ford directed that an airfield aptly named Ford Field be constructed in 1924, in the outskirts of Detroit. The facility was complete with paved runways, ticket counters, waiting rooms, and a hotel, the *Dearborn Inn*, built specifically for air travelers. Until then airfields had been nothing more than muddy, rutted treeless fields. Ford Field was a model for future air travel.

Ford began production of the Ford Tri-motor aircraft designed to carry up to twenty seven passengers. For the first time, a commercial airline began regularly scheduled passenger flights using an all metal airplane. Initial flights were between Detroit and Cleveland; then flights to Chicago were added. By 1933, Ford's experiment in aviation ended with the Great Depression and several millions in losses. However, during the mid-to- late '20s, the Dearborn airport made an impression on a young woman named Mildred Doran who was attending a nearby college.

2

MILDRED DORAN

Mildred Doran was the second oldest of the four children of William and Minnie Lee Doran who emigrated from Ontario, Canada, to Flint, Michigan, when Mildred was three. Flint was a bustling General Motors town offering good jobs at a fair pay. At the time, both Buick and Chevrolet automobiles were manufactured there. Mildred had two brothers and a sister. When she was only fourteen, her mother died of tuberculosis. Her father, for reasons not clear, was not able to take full responsibility for the children. As the oldest daughter, Mildred took on the duties of looking after her two younger siblings with the assistance of an Aunt and Uncle, and other members of her family.

The family lived on the south side of Flint in a working class neighborhood a little over a mile from the downtown center of the city. Most of Flint's population worked at the auto plants, Chevrolet, Buick, or Fisher Body. The wooden two story bungalows of the neighborhood were modest and tidy. Most had a front porch, shaded by the trees that lined the streets. Behind the homes, a dirt alley running between parallel streets, accommodated garages. Jobs were plentiful, and each family kept a car in the garage. The neighborhood kids played baseball and tag in the alleys. In the summer, it was swimming at Tread Lake on the other side of Saginaw Street. From this neighborhood, Mildred was close enough to walk to grade school and high school.

An honors graduate of Flint Central High School, she went on to Ypsilanti Normal College, now Eastern Michigan University, and graduated with a teaching degree at age twenty. She was a member of the Alpha Sigma Tau sorority.

While still in college, at the age of nineteen, Mildred attended an air show at Flint's Lincoln Airfield. As she walked by, one of the "gypsies" asked Mildred and a girlfriend if they would like to go up for a spin. Mildred hoped someone would tell her "no." Mildred said of that encounter, "I really did not want to go up at all that first time… I could have just prayed for someone to forbid me to go, [so] I wouldn't look a coward before the courageous girlfriend who was with me. But the owner of the field not only let me—he urged me; and what could I do?" The two girls talked over the offer for a moment and then agreed. Mildred said of that experience, "I am certainly nothing of a hoyden, and yet the first time I went up, I wasn't a bit afraid." That experience was a thrill, and Mildred knew from that moment that she wanted to be a flier.

". . . After we began rushing through the air, twisting and turning, it seemed like some particularly glorious ride on some miraculously smooth highway." She had caught the "air bug."

After graduation, Mildred began teaching fifth grade at Caro, Michigan, north of Flint. Her yearly salary was $1,200. She also taught Sunday school at Lakeview Methodist-Episcopal Church in Flint. In 1927, she was twenty-two years old.

Mildred Doran, five foot two, with a trim figure, had an olive complexion, wide set blue-grey eyes, bobbed wavy brown hair, and a captivating smile. In various newspaper accounts, her eyes are described as blue, blue-gray, dark blue, hazel, and brown. Obviously, not all of those descriptions are correct. It provides insight into the news reporting of the period. I assume her eyes were blue gray just as her sister Helen's were. One report referred to her "endearing accent" (perhaps Canadian?). During her many interviews with reporters, she was poised and open about her love of flying. Her straightforward style of expressing herself was probably considered bold in the '20s. In addition to flying, Mildred was an accomplished equestrienne and swimmer. It was common in news stories to describe her as "pretty" or "attractive."

While still in college, she began to "hang out" at the Lincoln Airfield at the corner of South Saginaw and Maple Streets in Grand Blanc Township just south of Flint. The airfield was owned by Lincoln Petroleum Company, a very successful enterprise that operated gas stations in central Michigan. Lincoln Oil was owned by William Malloska, who was both the employer and a friend of Mildred's father. Malloska had "kind of" adopted Mildred and helped her with her college expenses. It's doubtful whether Malloska himself was a pilot, but he did fly with a couple of pilots, Augie Pedlar, known for his wing walking stunts, and Eyer "Slonnie" Sloniger, a pilot who had flown for the Army in France during the War. Pedlar and Sloniger were associated with Lincoln Standard Aircraft Company in Lincoln, Nebraska. The similarity of the names of the two companies is happenstance. Occasionally, the Lincoln Standard pilots would pick up Mildred in Dearborn and fly back to Flint for the weekend. Mildred spent much of her spare time at Lincoln airfield, where she would bring everyone up to date on the latest in flying news. There was plenty to talk about. During this period, Mildred was permitted to occasionally take the controls of a plane during short flights around Flint, but she did not have enough flying time to qualify for a license.

3
THE DOLE TRANSPACIFIC AIR RACE

When the Lindbergh triumph of May 21, 1927, hit the headlines, James Dole, the "Hawaiian Pineapple King" was in San Francisco on business. Dole was already an aviation enthusiast and was responsible for Rodgers Airfield under construction at the time in Honolulu. Dole's fondest dream was to entice Lindbergh to fly to Hawaii. It would be a public relations triumph for Hawaii.

Dole was a successful entrepreneur in growing and canning pineapple. He started with sixty acres in 1899, and had more than 200,000 acres under cultivation by 1922. He helped develop a mechanical peeling machine which greatly increased his revenues. Until that machine was put into production, farms were limited by the number of farm hands one could find. His advertising campaign in magazines is often cited as the first to go nationwide in the U.S. Americans loved his canned pineapple. The ad beginning with the words, "You can thank 'Jim' Dole for canned Pineapple Juice!" was seen in almost every magazine.

James Dole was a wealthy man. He envisioned the time when Hawaii would be a major tourist destination. When that day came, the islands, and his holdings, would become even more valuable.

On May 25, 1927, just four days after Lindbergh's landing in Paris, Dole wired the *Honolulu Star Bulletin*. "James D. Dole, believing that Charles A. Lindbergh's extraordinary feat in crossing the Atlantic is the forerunner of eventual transpacific air transportation, offers $25,000 to the first flyer, and $10,000 to the second flyer, to cross from the North American continent to Honolulu in a nonstop flight, within one year beginning August 12, 1927." In today's dollars, the $25,000 prize is equivalent to about $300,000. A flight to Hawaii from California is just over 2,400 miles, virtually all of it above the Pacific Ocean, which, of course, is devoid of any visible navigation aids.

The following day, Dole spoke more in the *Star Bulletin* about the reasons behind his out-of-the-blue announcement. "Following Lindbergh's flight, it seemed obvious that a flight from the mainland to Hawaii should be next in order and that definite action to encourage such a flight would be appropriate. . . . .This prize offer coming from the islands encourages the development of something that is bound to become of great importance to all of us in the islands as well as to the world at large."

"It is natural for all of us in Hawaii, as well as all the followers of aviation the world over, to hope that this contest may be doubly successful, first that it may cost no brave man either life or limb, and second, that the continent and Hawaii may be linked by airplane."

The "Dole Derby" as it was called by the news media, set off a flurry of activity across the country. Dole had hoped to lure Lindbergh into the race, but Lindy was not biting. However, there was a profusion of other pilots dreaming of becoming a world famous celebrity like Lindbergh. They were ready to enter the race.

A cute ditty based on a nursery rhyme appeared in the San Francisco *Examiner*:

"Mister Dole is an airy old soul,

And an airy old soul is he.

He calls for a race from this here place,

To the place where the islands be."

News reporters frantically sought interviews with the prominent fliers of the day to discover which of them would be going after Dole's prize. Clarence Chamberlain and Commander Richard Byrd in New York indicated an interest. Hawaiian pilot, Martin Jensen, announced he had been planning for months to make the flight from California back to

Honolulu. Louis Berliot, Jr., son of a pioneer French aviator announced plans to enter a French plane. And the St. Louis backers, who had financed Lindbergh's Paris flight, announced that they would fund him again, if he entered the transpacific race. An Oklahoma woman wrote Dole to announce her intention to enter the race. The money was her motivation. As she states, "I am no aviator, but I sure can learn and take a chance for the prize."

A group of businessmen and sportsmen headed by Grant E. Dodge of Los Angeles declared on June 8, two weeks after Dole's announcement, plans for a race from that city to Honolulu and were seeking subscriptions from the public to offer $50,000 in prize money. Los Angeles wanted to upstage San Francisco in aviation promotion. Those plans failed for lack of sponsors and the Dole Race was on.

William F. Malloska, the owner of Lincoln Petroleum Co. (a/k/a Lincoln Oils) in Michigan was a successful entrepreneur. Malloska had begun his career as a carnival man, but made his fortune in the gasoline business. In 1927, he owned about seventy service stations in Central Michigan. He was well known around Flint as the generous sponsor of baseball and bowling teams. In a 1927 decision, the Michigan Supreme Court ruled against Lincoln Oil. Malloska had come up with a marketing ploy in which customers of his gas stations were given a coupon each time they purchased a tank full of gasoline. Every month Lincoln Oil held a drawing and the customer with the winning coupon was awarded a new automobile. The Court found the promotion was an illegal lottery and ordered that the drawings be discontinued. The case gives insight into Malloska, the promoter. He saw the Dole Race as another marketing scheme.

Mildred Doran overheard Malloska discussing his interest in entering the Dole Race. She asked, "Can I go?" She wanted to be the first woman to fly over the Pacific. We do not know how persuasive Mildred had to be to obtain Malloska's support, but she was irrepressible. He agreed to buy a plane and enter it in the Dole Transpacific Flight. Malloska agreed to pay all expenses. Mildred says, "I was so thrilled, I jumped into my car and dashed all over town." Malloska soon made sure his plans to enter the race with Mildred were known by the news media.

Newspapers throughout Michigan, where Mildred taught school all sought interviews with Mildred. The most frequent question was whether she was worried about the safety of the flight. Mildred professed no fear. "I am not a bit afraid," was her reply. She added, "I feel sure we will win, but if we don't—well life is a game of chance anyway." Later, she put it another way: "I know we will do it. I would not go if I had any misgivings about it. Life is sweet to me and I am not courting an early death…I don't consider the prospective flight as tempting fate or taking a great chance."

Mildred went on to say, "I have always traveled the well-worn road over which the majority of girls reach maturity. I never was given to wild adventure, but was rather quiet and kept out of the limelight. But the Lindbergh flight seemed to unleash something in me. . . ."

James Dole made his announcement without a lot of deliberation. Specifics were yet to be spelled out. Dole turned over the details of the race to the National Aeronautical Association, a nationwide club of aviation enthusiasts. The Association formed a race committee, with members in California and Hawaii. The California group was headed by Frank Flynn, a flying ace from World War I. He and three others established a set of rules for the race. There had never been such an event so the "rules" were frequently amended as the start of the race approached. The initial decision was the selection of a site for the start of the race. Dole's wire to the *Star Bulletin* had only referred to the "North American continent."

In April, 1927, the City of Oakland purchased property at Bay Farm Island, on the shores of San Francisco Bay, and began constructing a new airfield. By summer, it was incomplete, but had an essential feature: an extraordinarily long 7,070-foot dirt runway. Planes taking off loaded with considerable quantities of fuel required long runways. In addition, unlike other West Coast sites, there were no nearby hills and usually no fog.

The period from Dole's announcement on May 25, until the race on August 12 was only eleven weeks. That was not much time considering that the contestants had to acquire aircraft and make necessary refinements. Aircraft manufactured at the time were not capable of flying such long distances "as is" so modifications such as higher capacity fuel tanks and navigational instruments were added.

Over thirty five inquiries were received by the Race Committee. Among them was an inquiry from someone expecting to build a dirigible to enter the race. Early on, the St. Petersburg *Evening Independent* named several "unofficial" entrants: Major Livingston Irving, Berkley; Charles Vance, Pacific Air Transport; Lt. Jess Widham and Ben Stern, Memphis; Nicholas Miscovich, San Francisco; Robert Hart, Jr., Altoona; Ernest Smith, Pacific Air Transport; Paul Redfern,

Atlanta; and Capt. F.J. Franklin, Los Angeles. Other than Irving, none of them actually participated. The Race Committee required all contestants to deposit a $100 entrance fee to winnow out the less serious. The Race Committee was astonished to receive eighteen entries. Discouraged by the Race Committee's attitude, the blimp never entered.

Entries closed on August 2, with the race set for August 12, 1927. The date of August 12th had two things going for it. There would be a full moon that night. Flying time would be more than twenty-four hours, so at least half of the flying would be after dark. In addition, it was the anniversary of May 12, 1898; the date Hawaii became a United States Territory.

Dole had not envisioned that more than one pilot would be ready to fly by August, thinking that it would take several months or even a year for adequate planes to be built and outfitted. It was not at all predictable that the first aviators to attempt preparations for a trans-Pacific flight would be successful. When it became obvious that several would attempt the flight at noon on the date selected to start the competition, rules as to the starting order had to be designed.

As it turned out, several entries could not compete because their financing fell short, or their airplanes were not delivered in time. Two other entrants were disqualified due to their inability to carry sufficient fuel for the flight. On the morning of the race, there were only nine contestants ready and willing to embark on the expedition.

In the days preceding the race, Clarence Young, head of the enforcement division of the newly formed Aeronautical Branch of the U.S. Department of Air Commerce became involved in the safety aspects of the race. His division would certify pilots and navigators, as well as the planes before they would be eligible to race. When he arrived at Oakland on Saturday, August 6, he found only one plane had arrived to be inspected: the *Pabco Pacific Flyer*.

Young decreed that pilots must have an "air transport rating" in order to compete. This rating necessitated at least two hundred hours of flying time. In theory, pilots kept a log in which they recorded their flying hours. Many, especially veterans of World War I, never bothered with a log book. Young personally interviewed the pilots as they arrived at Oakland and granted licenses on the spot to those who convinced him of their qualifications, including Augie Pedlar, *Miss Doran's* pilot.

In addition to a licensed pilot, each plane was required to have a qualified navigator approved by the Department of Commerce. Pilots generally navigated long distance flights using roads and railroad tracks to find their way. Unlike Lindbergh's flight, almost the entire distance of the flight to Hawaii is over water with no landmarks. A navigator was expected to keep the plane on the most direct course in order to save fuel. He would also be responsible for finding the Hawaiian Islands at the end of the journey. The long flight over water was too dangerous for a novice navigator. To qualify, prospective navigators were first given a paper and pencil test. Those that passed were taken on test flights around Oakland to demonstrate their proficiency.

A formula was developed to determine the amount of fuel that had to be on board, based on the fuel consumption of the individual plane flying at a normal cruising speed — with some extra in case of headwinds. For the great distance involved, extra fuel tanks were fabricated and placed in the wings or fuselage. Fuel/ Distance calculations were in their infancy. Officials in charge had no personal experience in flying distances of over 2,000 miles nonstop. Because of the danger involved in taking off and flying with a full load of fuel, calculations were based on the plane flying with only a half of its fuel capacity.

The racers were required to carry life boats, sails, flares, and emergency rations. Wing lights and tail lights were installed to satisfy new Department of Air Commerce regulations never before enforced. Young felt that the planes, all following the same course all night to Hawaii, needed lights for safety. As it turned out none ever flew in sight of any other entrant. Most of the planes had to install lights after arriving at Oakland. There were no requirements as to carrying a radio and only a few could afford that extra expense.

On Monday, August 8, a drawing was held among the fifteen planes expected to show up for the race to determine the order of takeoff. The plan was to have the fifteen planes takeoff, one at a time, at two-minute intervals. Numbers were pulled out of a waste basket in the offices of Capt. C.W. Saunders in the Matson Building in downtown San Francisco. Only five of the entries were represented personally. Mildred Doran dressed in a tailored suit and cloche hat, reached in and extracted number four.

By the Wednesday before the scheduled start on Friday, August 12, only four of the planes and crew had completed their tests and inspections. Marty Jensen's plane, the *Aloha,* was wheeled out of its San Francisco factory on Tuesday, August 9, with its bright yellow paint barely dry. On the following day, *Aloha* made its maiden flight from San Francisco to Oakland Field. Others were contemplating whether to announce their withdrawal. Four were hoping weather would permit them to reach Oakland by Thursday night in time for the start on Friday. Obviously, the rush of entrants this close to the Friday start did not allow sufficient time for checking pilot and navigator qualifications or inspection of the aircraft. The race committee held an emergency meeting starting in the early evening, and lasting until after midnight. In the wee hours of Thursday, August 11, they decided to postpone the start of the race until the following Tuesday, August 16, 1927. The eleven weeks between the announcement of the race and August 12, turned out to be just too brief a time for most of the entrants to acquire a suitable plane and to pass the inspections. Despite these obstacles the race was still acclaimed in the press. For example, the *San Francisco Bulletin* declared, "The greatest race in the history of the world—San Francisco to Honolulu by air! The biggest sporting event ever since the Roman gladiators went into the arena—a score of men and one woman playing on a tenuous line between life and death!"

4

Miss Doran Heads To California

Eyer "Slonnie" Sloniger was one of ten siblings raised in western Nebraska. He was born in a sod house in western Nebraska. All ten of the Sloniger children attended college, but Eyer was the only one not to graduate. By the middle of his first year in college, in 1917, Sloniger joined the Army and volunteered for pilot training. In March 1918, Sloniger took his first solo flight in a Jenny at Kelly Field in Texas. Slonnie said, "I loved aerobatics and never understood why so many guys were scared to death of them." Slonnie understood that flying by the seat of your pants was "a bunch of hooey." He devised a rudimentary "instrument." He hung a heavy steel nut on a piece of string and attached the string to the cowling over the instrument panel. When it hung straight down, his plane was level.

After flight training, the Army sent Slonnie to France to join the War. Then, before seeing action, the Armistice was signed in November 1918. A career in flying was born.

Sloniger immediately got together with about ten other Nebraska pilots returning from the War, and began taking passengers for twenty five dollar rides. At first, the group of veterans had as much business as it could handle. It was "fill up and haul some more." Slonnie would take his customers up for a ride which included barrel rolls, loops, and figure eights. After a few months, the number of passengers declined. The ten pilots dwindled to three. The public demand for air rides diminished. Tickets to fly were reduced to five dollars. Fewer customers showed up to fly.

Nebraska Aircraft in Lincoln, Nebraska, landed a contract to deliver air mail in Mexico. Slonnie accepted the offer of steady work to fly mail down there. The contract was terminated after only one year. But, a new Mexican flying job came up. The oil field workers at Tampico demanded to be paid in gold and silver coins. Trucking the money from Texas was dangerous. As Slonnie stated, "There was no shortage of people out to grab payroll shipments. So the oil companies turned to air delivery,"

In 1921, Slonnie returned to Nebraska, saying he was sick of Mexico. Ray Page had taken over Nebraska Aircraft and re-named it Lincoln Standard Aircraft. Planes were assembled in a field as there were no hangars. Slonnie would test the new planes just before customers picked them up. Slonnie said he flew "damn near every plane" manufactured by Lincoln Standard. It was during a test that Slonnie had his only crash landing. He was flying upside down and the plane went into a spiral. Sloniger leveled it out just before it hit the ground. He was unhurt. The plane was a total loss.

As a way to earn publicity for its planes, as well as to earn additional money, Lincoln Standard started a traveling air show which barnstormed around the Midwest. It was advertised as an "Air Circus." Posters proclaimed the circus was "more daring, more thrilling and more elaborate." Sloniger joined the rough and tumble life of a circus pilot. There were air races, stunt flying, wing walking, and parachute jumping demonstrations. After the show, the pilots would make added money by taking passengers up for stunt flying demonstrations. While flying in the traveling Air Circus, Sloniger first met Augie Pedlar. Pedlar was a wing walker in the show.

The story of how Sloniger got his first plane gives insight into the life of a barnstormer. Ray Page had obtained a Fokker D-VII used by Germany in the War. At each performance of the air circus, crowds gathered around to scrutinize the Fokker. Sloniger said it was the best handling plane he had flown. He wanted it.

One day the opportunity presented itself. After a show in St. Joseph, Missouri, Slonnie found himself in a crap game which included Ray Page's girlfriend. Slonnie made twelve straight passes, doubling after each one, winning $1,500. The girl couldn't pay. So Slonnie approached Page and rolled once more at double or nothing with the Fokker priced at $3,000. Slonnie rolled another pass and the Fokker was his.

The Lincoln Standard bunch made their way to Michigan with the Flying Circus in 1926. They settled into Flint in Central Michigan as their center of operations. The Flying Circus found large, enthusiastic audiences throughout Michigan. Customers were lined up "in fours" to buy tickets. In Flint, Slonnie met Bill Malloska. As Slonnie recalls, "...Bill Malloska, who had cut-rate gas stations all over town, was giving free gasoline to everybody...[You] wanted a base like Flint where General Motors was. The well paid factory men were good customers."

Malloska proposed that the words, "Lincoln Oils" be painted on the side of the circus planes in exchange for free gas and oil. Since the name "Lincoln" fit right in with Lincoln Standard, and the paint did not add weight, Slonnie declared, "Paint it on."

In 1927, when Lindbergh succeeded, Slonnie thought about the money involved in winning the Dole Derby. In discussions with Malloska, Slonnie suggested using a Lincoln

Standard plane with a Wright Whirlwind engine, the same engine Lindbergh used in the *Spirit of St. Louis.*

When William Malloska set out to find a plane for the Dole Race, he initially placed an order for a "giant" monoplane with Ray Page of the Lincoln Standard Aircraft Company in Lincoln, Nebraska. Newspaper reports stated that aircraft "experts" were working day and night to complete the giant plane in time for the race, building it with extra capacity fuel tanks. The Wright Whirlwind J-5 engine, a nine cylinder radial engine, was manufactured by Wright Aeronautical Company in Dayton, Ohio, which was the corporate name of the aircraft engine company started by the Wright Brothers. On completion the plane was to be delivered to the Lincoln Airfield in Flint and housed in a specially built hanger large enough to accommodate the massive plane. However, it soon became clear that Malloska's order was not going to be completed in time and Malloska turned to a manufacturer closer to home.

Malloska purchased a Buhl Air Sedan for the Dole race for $12,700. He later claimed he spent $30,000 (about $360,000 today) for the plane and all its equipment. Buhl Stamping Company was headquartered in Detroit and had been in the metal-forming business for about a hundred years. The ornate, art deco style Buhl Building in downtown Detroit, which served as Buhl's headquarters, is still in use today. Buhl began manufacturing airplanes in 1925 at a plant in Marysville, Michigan, about seventy miles east of Flint, near the shores of Lake Huron. Buhl was known for the quality of its planes and was chosen by many for speed and reliability races in the late '20s and early '30s. However, in the '30s the Great Depression did in Buhl Aircraft Co. Its doors closed in February 1933.

Descriptions of the Buhl aircraft stated, "the plane will be equipped with a 'Wright Whirlwind' engine similar to that used by Colonel Charles A. Lindbergh in his flight to Paris and will burn about three hundred gallons of gasoline." The plane was expected to be delivered in about two weeks. Sloniger recalls that the Buhl was about the "only thing we could get at the time."

The Buhl Air Sedan was the only two-winged aircraft to start the Dole Race. Sloniger described it as a "sesquiplane really, not a biplane," as the two wings were uneven in length with the lower wing being shorter. Although the correct term is sesquiplane, it was almost always referred to as a biplane. It was thirty feet long with a forty foot wingspan for its upper wing. It was powered by a three hundred horsepower engine, and had a fully enclosed cockpit with seats in front for the pilot and navigator. Another seat was placed behind the left wing with a window for Mildred. The aircraft was painted red, white, and blue, with a red nose and wings, white fuselage, and blue tail. A large Lincoln Oils logo was painted on each side just behind the wings.

Both Augie Pedlar and Slonie Sloniger wanted to pilot the plane to Hawaii. There was not room for both because there had to be a qualified navigator on board, and neither man could pass the test. As Sloniger recalls, "Augie Pedlar had been my wing walker out of Lincoln and those fellows usually rode along when we moved from show to show. They'd hook up controls in the front cockpit and you'd let them practice. He finally picked up enough flying time to go out west and get a flying job. But, he cracked up pretty badly and was very crippled with one leg turned way under. An awful nice guy though, who'd heard about my Buhl and was hanging around Flint. . ."

The two airmen campaigned to be the pilot chosen for the race. Neither would back down and the dispute was heading for an ugly finish. Finally, their Flying Circus supervisor, "Pop" Morris suggested the matter be decided by the flip of a coin. Sloniger protested, "Flip Hell! That's my airplane!" In rebuttal, Pedlar argued that Sloniger would fly the rest of his life, but he (Pedlar) was handicapped by his leg injury. "[He] couldn't make enough money flying but he might get something out of this Dole Race to start a little business." Morris leaked it to the press that there would be a coin flip and a meeting was held in the offices of the *Flint Journal*. Under pressure, Sloniger finally relented and flipped his "lucky" fifty-cent piece. Pedlar called heads and won. Afterward, Slonnie gave the coin to his wife for her charm bracelet.

In newspaper interviews at the time, Pedlar disputes Sloniger's claim that he had meager flying experience. Ironically, Pedlar, twenty four, caught the flying bug at age fifteen when he saw a sign in a Long Beach, California movie theatre offering money to the first pilot to fly to Hawaii. He told reporters, "I never forgot that sign." It is unclear where Pedlar did his flying in California. In the air circus, he did wing walking, hung from the landing gear, and performed "other death defying stunts." Pedlar claimed to have trained to be a pilot under the tutelage of Ray Page, who ran a flying school at Lincoln, Nebraska. Page was the teacher who had trained Charles Lindbergh in his early flying days. Pedlar, like Sloniger also flew air mail in Mexico for Nebraska Standard. Pedlar was proud to be a barnstormer, saying that he would trust any barnstormer over a "dude" pilot. Pedlar claimed to have over three thousand hours of flying time by 1927. Sloniger doubted it was even the minimum two hundred.

The Air Sedan was delivered to Lincoln Field in Flint. A few days later, on Saturday, July 9, a christening ceremony was held. Several hundred spectators were present along with the Mayor of Flint. Dignitaries spoke and a band played. Also present were Mildred's father, her brothers, William, a law school student, and Floyd, aged fifteen, and her little twelve-year old sister, Helen Doran.

The plane was christened with a bottle of ginger ale broken across its propeller since it was during Prohibition (1920-1933) and alcoholic beverages were outlawed. The plane's name that had been a secret until the ceremony was dramatically announced, *Miss Doran*. The name was written in script on the rear of the fuselage, and on the engine.

An article in the *Flint Journal* claimed that little sister, Helen, had been angry with Mildred for two months because she had wanted to go on the flight too, but Mildred would not hear of it. Mildred once said, "I gave everybody a shock when I first suggested the idea [of flying to Hawaii] and apparently they haven't gotten over it. If it had been my sister, Helen, everyone said they wouldn't have been surprised. Helen is twelve and a tomboy, and has done quite a bit of flying herself."

The fact that a woman was going to be involved in the race was a novelty and the decision to allow her to participate was questioned. Malloska said, "I have all the confidence in the world in Miss Doran. I know of the hardships she endured after the death of her mother. She was able to go to school only a half of each day, worked in a telephone office the other half, and in addition took care of all the housework and was a mother to her younger sister and a young brother. She will go through this flight just as she has gone through other hardships."

Mildred said, "I have had considerable experience flying. I feel perfectly at home in an airplane, and do not fear the trip at all." She reportedly felt that she could take over the controls if it became necessary. She said, "I am not a licensed pilot and therefore, cannot fly across state lines. However, I have driven a plane considerably and am confident I could pass the examination. I won't handle the plane unless an emergency occurs, and I am confident that I could handle it then all right."

In an article written by Mildred for the *Detroit Times* she said, "Since last spring, it has been my ambition to be the first person to fly across the Pacific, and . . . I want to be the first woman. [I]s there any reason why a woman can't do anything a man can do in aviation? Women certainly have the courage and tenacity required for long flights. . . ."

Mildred's father William Doran was questioned by a reporter as to whether it was wise to permit his daughter to go on the trip. His answer seems a bit odd. He replied, "Sir, my daughter is not yellow, nor is there one member in the family with a yellow streak; the Doran name stands for courage."

When asked by *San Francisco Call-Bulletin* news reporter Mary Mattison about Mildred's courage, Augie Pedlar, *Miss Doran's* pilot recounted an incident that took place at an air exhibition in Lansing. Sly Sloniger, standing in front of his two-seater, announced to the surging crowd surrounding him that he was going to take his plane up to seven thousand feet, turn off the engine, and bring her right back to where they were standing without power. "Who would like to go up with me?" he called out. According to Pedlar, the crowd nervously backed away. That left one diminutive figure standing in front. Sly laughed and said, "All right. Come on, Mil."

From Flint, the plane went back Marysville, Michigan, on July 11, to the Buhl factory for "one more navigating instrument." Another news article indicated the trip to the factory was to adjust the lower wing. Then *Miss Doran* was to fly to Selfridge Field in Mt. Clemens, Michigan, on the shores of Lake St. Clair, where its barograph, used to show altitude by air pressure, was to be sealed by government officials. Next, it was over to Ford Field in Dearborn for the gas tanks to be fully loaded for the trip to California.

At the time, Mildred planned to take her three-week old Great Dane named, Honolulu, on the flight to Hawaii. The puppy had been a gift from Bill Malloska. It was a pure bred dog from Ali Baba Kajaba and Lady Von Strohelm, costing a reported $2,500. It was a brother to a pup purchased by Edsel Ford, Henry's son. *Honolulu* made it to Oakland, but in the end the dog did not make the flight from Oakland due to illness.

On Tuesday, July 12, Pedlar took the plane up for the short flight from Selfridge to Ford Field. At seven thousand feet the plane was hit by a strong northeast gale streaking across the lower peninsula of Michigan. The plane lurched. Suddenly, the cabin door blew open. Mildred and William Malloska, as passengers, almost fell out. The plane shuddered Pedlar surveyed the situation and saw the fast moving storm about to overtake them. He decided the storm was approaching too fast and turned back to Selfridge just managing to land and get the plane into a hangar as the storm hit. After getting the plane safely in its berth, Pedlar declared he had put the plane through its paces. In order to beat the storm, he had the plane at one-hundred-and-fifty miles per hour, thirty miles per hour above its normal cruising speed of one-hundred-and-twenty. Pedlar, aged twenty-four, said this was the first time in his life he had ever been forced to turn back due to weather.

Bill Malloska described the incident for reporters. "I admit I was very much thrilled by the affair. The door on my side of the cabin opened as we shot upward. I grabbed a part of the framework with one hand, and that alone saved me from falling from the ship. After I had slammed the door shut, I threw the dog and traveling bags to the forward part of the cabin. At that moment Miss Doran was hurled across my body. We were shaken about in that plane for five minutes as popcorn is shaken in a popper. When Pedlar made his turn, I knew he intended to race the storm so I straightened out myself and Miss Doran. When we arrived a Selfridge Field, I realized for the first time that Pedlar was one of the best pilots I had ever known. He wasn't the least bit excited yet he did what was necessary when that wind shot us into half stall. "Mildred sympathized with Honolulu: "He whined pitifully until we landed."

The storm that raced across the lower peninsula of Michigan that day was responsible for three deaths, twelve injuries and damages worth more than one million dollars to buildings, power lines, and trees. Mildred was not the least daunted by the experience, and was confident in the man at the controls (Pedlar). She commented that Augie really "had her (the plane) going."

As a result, the flight to Oakland would begin at Selfridge Field rather than Ford Field. The runway at Lincoln Field in Flint was considered too short and bumpy to accommodate the plane fully loaded for a cross-country trip. The *Miss Doran* set off for California with Augie Pedlar at the controls and with William Malloska and Mildred Doran as his passengers, as well as the dog, Honolulu.

Sloniger recommended to Pedlar that he take the plane nonstop to California to test the equipment and to gain experience in piloting for long periods. Sloniger doubted that Pedlar had enough experience, and many doubted that the Pacific Ocean was the best place to practice one's first long distance instrument flight. Sloniger commented years later, "Pedlar didn't give the plane a fair shot. I told him to fill it up with gas and fly clear to the coast in one hop. Get a feel for what it would do and how much fuel it needed. But, Bill

Malloska, with his carnival ideas, rode out to California in *Miss Doran* and they had to keep landing for some more publicity."

Pedlar told the *Detroit Times*, "We expect to take the flight to Long Beach is easy stages. Our present plans call for stops at Chicago, Illinois; Tulsa, Oklahoma; Port Arthur; Texas; Long Beach, California; and our final goal of Honolulu, Hawaii. I am ready for the trip and nothing can stop me. The ship is a wonderful plane.... I am staking my life upon my belief in the ship."

From Flint, the three made their way to Chicago, and then St. Louis. The trip was delayed for a day in St. Louis due to bad weather. A postcard sent air mail, of course, to little sister, Helen, from St. Louis on July 18, 1927, says: *"Dear Sis & Family. Spent last nite in Chicago. Headed for St. Louis and Tulsa today. Chicago is very pretty from above. Much Love, Mil."*

On Monday, July 18, the *Miss Doran* went from St. Louis to Tulsa for a fuel stop and on to Ft. Worth, Texas where they landed in the midst of an air show held to dedicate Meacham Field. A crowd of 15,000 gathered to watch aerial demonstrations by forty-six Army planes as well as twenty commercial planes, including parachuting demonstrations, and other stunts. A traffic jam ensued just after the *Miss Doran's* arrival at 10:15 am. It seemed everyone wanted a close up view of Mildred.

While in Ft. Worth, Mildred was interviewed. She told the reporter, "I like to fly and have been interested in aviation for several years, but I don't intend to give up teaching for it. I like teaching too well. What I am attempting to do will be nothing out of the ordinary in a few years. Flying an airplane in a few years will be just as common as driving an automobile."

The following day they took off for El Paso, Texas, and on to Tucson, Arizona. The airport log from Davis-Monthan Field in Tucson shows that the three of them arrived there on July 19, 1927 from El Paso, Texas, for a short refueling stop. From Tucson they proceeded to Long Beach, California. Upon landing, a bit deaf from the noise of the flight, Mildred was quoted as saying, "I have been flying for three years and love the air."

Pedlar originally planned to take off for Hawaii from Long Beach. It was his hometown and the place where he first started to fly at the age of fifteen. The original Dole announcement did not specify a takeoff location other than "the North American Continent." It was three weeks and a day before the August 12 race. Pedlar gave a speech to a well-attended meeting of the Long Beach chapter of the National Aeronautical Association and the Chamber of Commerce in which he announced his intentions to take off from Long Beach on condition that the runway be extended to two miles from its current one mile. As an alternative, he said that he would take off from the seven thousand foot City of Oakland runway in the event that the Long Beach runway was not extended (it wasn't). News reports of the Long Beach event mentioned that Mildred appeared tired from the cross-country flight, but remained confident. "We will be the first ones off on August 12. I'm really tickled to pieces to be here, and I haven't the slightest misgiving about the coming jaunt to Hawaii. I've always wanted to do something different and to be the first woman to do it. Even if Augie and I don't pick up first money, I will still be the first woman to make a trip like that."

The following week, the *Miss Doran* was taken to San Diego. From the San Diego Hotel, Mildred sent a letter to the family back in Flint:

"San Diego, Calif. July 31, 1927. *Dear Folks, Just finished writing seven letters, so while I'm still in the mood, I won't forget you.*

"*Everything is going fine with us. Lincoln left last nite for the Hawaiian Islands. He will be wearing a grass skirt and doing the Hula Hula dance when we arrive. The western cities have treated us in a royal fashion. Lots of things given in our honor.*

A Mr. Doran called on me from California. He was born in Flint, but left 40 years ago. He is about 60 years old. His sister's name is Miss Lane and is related to the Doran in the Drug Store. He is a millionaire. She drives a Lincoln with a chauffer (sic.) and he a Packard. They drove about 150 miles to see me and stayed a couple of days to entertain me. They are very aristocratic and dignified. I never acted so dignified in my life as when I was with them. They took a great liking to me. They haven't any children of their own. Called me their darling Mildred. Promised to stay a week with them when I came back. Lincoln wouldn't meet them for was afraid of shocking the old lady.

Had a long letter from Aunt Bessie and Elmer [Gonsler].

We are leaving for San Francisco next week. I'm anxious to get started on our trip. I have been studying navigation. I, and the other navigator will both navigate. I'll send you all a grass skirt from Honolulu. Guess you could do quite a stroke of farming them.

Now dear folks, my hand is getting tired of pushing this pen so will send you another edition tomorrow.

Heaps love, Mildred"

The person referred to as "Lincoln" must be William Malloska, the owner of Lincoln Petroleum Corp. Why the "old lady" would be "shocked" is left to the imagination.

On August 4, the San Diego Hawaiian Club sponsored a dinner at the Golden Lion Tavern to honor Mildred, "the flying schoolmarm." In its announcement, the Club said, "Believing that having given a send off to the first man who made an attempt to fly across the Pacific [John Rodgers], it is only fitting the same tribute should be paid to the first woman making the same attempt." Invitations were issued to members, and the general public. Over sixty San Diegans attended. Mildred addressed the crowd. She vowed not to take a life preserver. "We are not going down out there; we are going to stay up until we reach land."

From the start of the journey from Flint to Oakland, Mildred received hundreds of letters from fans, including mash notes from boys across the country. News stories about her and her attempt to fly to Hawaii appeared nationally as well as in every city along her route. When she arrived in California she mentioned that she would be wearing several fraternity pins to Hawaii. In photos taken prior to the race, the number of pins increased from four to six by race day.

On the last leg of its flight to Oakland from San Diego, the *Miss Doran* was forced to land in a wheat field in Mendota, California, eighty miles short of their destination. According to some newspaper reports, a spark plug blew up and gasoline sprayed out of the resulting aperture onto the engine. The hot engine ignited the fuel and Pedlar, Mildred, and Manley

Lawing, a Navy navigator recruited at North Island in San Diego, had to fight the flames with extinguishers, while searching for a suitable place to land. The newspapers painted a picture of Augie Pedlar guiding the plane's controls with one hand, and with the other, reaching out the window and showering the flames with a fire extinguisher. The Doran family scrapbook, containing these articles, has a hand written notation. It reads simply, *"BUNK."* In a different account, Mildred describes the event as less dramatic. "...(N)earing Mendota, the engine started missing. Augie thought it best to land and took it [the controls] over. He made a perfect landing in a wheat field, but the field was rough and the left wheel and landing gear got caught in a rut and were torn off." Augie Pedlar and Manley Lawing were stuck in Mendota changing spark plugs and hiring a welder to repair a broken wheel strut.

Miss Doran was not alone in its problems. Bennett Griffin's *Oklahoma* was forced down in Amboy, and again in Santa Monica, California, for repairs to its exhaust pipe. Otto Shafter and William Tremaine were forced to land Covell's entry, *Humming Bird,* at Escondido on the short flight from Los Angeles to San Diego due to a faulty oil line. Jack Frost's *Golden Eagle* damaged a wing and landing gear as it came back from a test flight in San Diego.

Malloska departed San Diego by ship for Hawaii. Upon his arrival he reported that he did not have a clean shirt. A small newspaper article told the story. "John D. (sic) Malloska, of Flint, Mich. . . .arrived in Honolulu today by steamer to find himself without a clean shirt. His laundry packages had been switched in San Diego. He had the one belonging to Miss Doran. 'Gosh, Mildred's got to make this flight now,' exclaimed Malloska."

The Oakland *Tribune* sent a car to Mendota to pick up Mildred Doran and take her to San Francisco. To scoop the other papers, the car had several *Tribune* reporters who interviewed her along the way. In the resulting articles, she recounted that she was a teacher from Flint and that Bill Malloska was her godfather. She joked that Augie was having some difficulty with the repairs to her plane. It seems that after they left San Diego, they had jettisoned their tools which were rattling around on the floor creating a noisy nuisance.

When asked about her pilot, Augie Pedlar, Mildred recounted that Slonnie Sloniger had once told her that Augie Pedlar was the one man that he knew who "knew no fear." She exclaimed, "I am so excited and thrilled and happy. I don't have words to express my feelings, and I'm looking forward so much to our arrival in Honolulu." The press loved the quotable Miss Doran and frequently called her the "Flying Schoolma'am, "Flying Schoolmarm", or the "Flying Teacher."

The following day, Mildred made her way to Oakland Field. As she arrived she was surrounded by reporters. One article was effusive: "Suddenly a small figure bends lithely, slips under the swaying fence, small white shoes resolutely challenging the ruinous brown dirt. It's Mildred Doran.

She wears a soft, sand colored traveling coat with a collar of ashes of roses, and a felt hat in the same shade. Under the coat is a simple white silk sports dress; on her hands are white fabric gloves with a small yellow border.

Mildred always wears gloves, an intrinsic neatness and trimness about her small figure made her a quaintly appealing picture when she folds those white gloved hands quietly and smiles at you from under the rose hat.

She's a pretty girl, Mildred Doran. A thousand city editors all over the United States must have beamed in satisfaction when they first saw her photographs, sent out from Flint, shortly after the announcement was made that she would fly the Dole race....

But although she photographs remarkably well, it's hard for a camera to suggest her wide hazel eyes and warm olive coloring, or the dimple that comes with her infrequent smiles."

Throughout the trip out west, Mildred emphasized the fact that women were equal to men. At Ft. Worth she declared, "There is no reason a woman cannot be just as good a pilot or flier as a man. There are many men who cannot fly without being ill. The only thing which bars women from any line of endeavor is lack of physical strength." In California she told reporters, "Why shouldn't a woman try it. There isn't a thing a man can do nowadays that a woman can't and so far as flying to Honolulu is concerned, I'm out to prove it. I want to be the first woman to make the flight and I'd like to see anything stop me now." At Long Beach she repeated her comment that women can fly just as easily as a man. She predicted, "There will be more women flying soon. When they learn how thrilling, comfortable, and clean this kind of traveling is they'll start in. That's the reason I am in a hurry to make this flight. I want to be the first woman to make a long non-stop flight."

Richard DuRose

5

John Rodgers Almost Makes It

Before the Dole Race, the dream of flying from California to Hawaii had been on the minds of flyers for years. Augie Pedlar had seen that poster in his movie theatre as a boy. At least one adventure novel for youth, *First Stop Honolulu,* in the Ted Scott series, had described how it could be done.

Two years earlier, in 1925, Naval Officer John Rodgers attempted to fly from San Francisco to Hawaii. Rodgers at the time was the Commander of the Naval Air Station at Pearl Harbor. His original plan was for three naval planes to make the flight. All three of the planes were built at the Naval Aircraft Factory in Philadelphia. The planes were "flying boats," two P-9s (sometimes referred to as PN-9's) and an experimental PB-1. One of the three was not ready on the day of the flight. Rodgers and another Navy plane took off from San Francisco in the late afternoon of August 31, 1925. The second plane had engine problems after about three hundred miles and turned back. Rodgers and a crew of four continued. Rodger's P-9 was equipped with three Packard 475-horsepower engines capable of flying at over eighty miles per hour.

Naval ships were stationed at two-hundred-mile intervals along his route belching black smoke by day and lighting the sky with searchlights at night as aids to navigation. The three Navy planes were rigged to land and take off on water and the ships closest to Hawaii were equipped to refuel if necessary.

Pre-flight calculations took into account the normal northwest to southeast trade winds, which saved fuel. The Navy assumed an additional fifteen miles per hour flying speed due to the winds. About half way Rodgers realized his fuel remaining was insufficient and he radioed for directions to the nearest refueling ship. There was a mix-up. Rodgers turned north when the ship was to his south. After searching unsuccessfully for the ship, Rodgers continued on. Ultimately, both his engines failed for lack of fuel and he glided toward the ocean. As he lowered his sea plane to the ocean, his radio went on the blink. The radio was powered by wind turbines and at the slow gliding speed they failed. Maneuvering for a dead stick landing, Rodgers sailed into a squall and successfully put down in the sea. Rodgers had flown 1,841 miles, at the time a new non-stop distance record, but well short of the Hawaiian Islands.

Rodgers and his crew were adrift in the Pacific Ocean. He had not broadcast an SOS, so there was no immediate chance of rescue.

Considering their predicament on that first day, spirits were high as they had landed safely. Rodgers had the men make preparations for a long sea voyage. Fabric from the wings was ripped off to make a sail and floor boards were torn out to make a rudder. Wood from the wings was broken up and burned in an attempt to distill fresh water. By the fourth day, all food was gone. On each succeeding day, the crew's spirits deteriorated. On the eighth day the men, exhausted by the sun and the sea, were losing hope. That night, they finally spotted the lights of Oahu fifty miles in the distance. A squall came up and brought much needed fresh water. Weary but uplifted, the men could picture the end of the ordeal. Although Oahu was in sight, Rodgers navigated the flying boat toward Kauai, which he calculated was an easier sail. As they approached the harbor at Kauai, the men waved sail cloth and set fires to draw attention. A Navy submarine spotted them, and came out of the harbor to tow the plane to safety. After nine days at sea. Rodgers and his crew were hailed as triumphant, even though they had fallen short of their goal by five-hundred-and-seventy miles and had been adrift for nine days. The fact that Rogers and his men had sailed for nine days and actually made it to the Hawaiian Islands was remarkable.

Ships had been dispatched to find Rodgers, but through a series of miscalculations, they never did. Navy analysts forecast that Rodgers would drift at four knots toward Hawaii, when he was actually drifting at three knots. Thus, the search proceeded ahead of the P-9 afloat in the Pacific. Navy patrols all but gave up the search after a week.

In recognition of his ability to navigate and his heroic leadership in saving the lives of his crew, a new airfield being constructed in 1927 at the Barber's Point Naval Air Station was named Rodgers Field. Rodgers Field was located on what is presently the Honolulu International Airport.

❻

THE ARMY AIR CORPS SUCCEEDS

The Army Air Corps wanted to complete a successful flight to Hawaii. No doubt it wanted to show up Rodger's flight for the Navy. Army Air Corps lieutenants, Lester Maitland and Albert Hegenberger were stationed at Dayton, Ohio's McCook Field. Maitland had been a flight instructor during World War I at the age of nineteen. After the War, Maitland was a member of the Army's racing team and was the first pilot to fly at more than two hundred miles per hour. Hegenberger also was a flight instructor during the War. In later life, he was known for his work in developing aircraft instrumentation. Their preparations for a flight to Hawaii began before Lindbergh's flight and Dole's announcement of a race.

McCook Field was the Army Air Corps headquarters for research and development. McCook Field is within a stone's throw from present day Wright Patterson Air Force Base, which continues to be a center of research for the Air Force.

A Fokker C-2-3 with a Wright 220 engine was modified for the flight. Although Fokker D-VIIs were used by Germany during the War, Anthony Fokker was a native of Holland. To satisfy the growing market in the U.S. after the War, Fokker built an aircraft manufacturing facility in New Jersey. The Army took delivery of a U.S. made Fokker.

The primary modification to the plane in Dayton was the addition of a longer wing. In preparation for the flight to Hawaii, pilot Maitland and navigator Hegenberger flew hundreds of miles in fog, clouds, and darkness to simulate flying over the Pacific Ocean. After months of training, they were ready, but the Dole Derby was still more than two months away. Their aircraft, *Bird of Paradise,* was geared up for the transpacific flight. The Army did not want to wait for the August date.

On June 15, 1927, Maitland and Hegenberger took off from Wright Field in Dayton and began the journey to Oakland. They intended to keep news of their mission quiet. But when they landed at Kelly Field near San Antonio, Texas, their fourth stop, they were greeted by a large crowd including reporters and cameramen. Coincidentally, Kelly Field was where Lindbergh had graduated from Advanced Flying School in 1925. The Army had leaked to the press information of their attempt at the Hawaii flight. This was exciting news! When asked, the crew said they were not interested in the prize money of the Dole Race, but wanted to make the flight to Hawaii solely for the advancement of aviation.

Along the journey to the West Coast, minor changes and repairs were made to *Bird of Paradise*. In Texas, a defective compass was repaired. At San Diego, an additional seventy-gallon fuel tank was added to bring the plane's fuel capacity to 1,120 gallons. At Oakland before takeoff Maitland and Hegenberger discarded their parachutes reasoning they would be useless over the vast Pacific Ocean.

On June 28, 1927, less than two months *before* the Dole Race, *Bird of Paradise* took off from the new airfield in Oakland and almost twenty-six hours later landed at Wheeler Field on Oahu. Theirs was the first successful transpacific flight to Hawaii.

While their flight was heralded as comparable to Lindbergh's, it was not without its difficulties. Within two hours, Maitland's two-way radio failed. Repairs were made, but after only thirty minutes, it failed again. One of the purposes of the flight was to test the long wave directional radio beacon from Maui. Without a radio, this goal was left unfulfilled. One engine sputtered temporarily due to oil getting into the carburetor. The induction compass failed. Hegenberger had to rely on magnetic compasses known to be unreliable at altitude. Later, another engine stopped due to icing. However, when Mailand lowered his altitude, the engine unfroze and resumed running smoothly.

The fact that Maitland could keep the tri-motor upright in spite of two engine failures is extraordinary. Unlike a one or two engine plane, the failure of one of three engines on a tri-motor renders the aircraft almost uncontrollable.

Adding to their woes, neither Maitland nor Hegenberger could locate the food that had been packed for them in Oakland. To their chagrin, it was found under their plotting board once they arrived in Honolulu. About 3:20 a.m., the crew of *Bird of Paradise* spotted the lighthouse at Kauai, seventy-five miles from Honolulu. A night landing in the rain, on the hilly terrain of Oahu, was risky. Instead, they began to circle at a slow speed. Finally, at dawn, they came in for a landing at Wheeler Field at slow speed bringing the plane in smoothly at 6:29 a.m. To the two happy fliers, breakfast was a very welcome meal.

Their flight was celebrated by Hawaii and the nation. Thousands of Hawaiians had gathered at the field in anticipation of their landing. A booming field gun salute welcomed them. The *New York Tribune* wrote, ". . .the cheering crowds at Honolulu see themselves emerging from the lonely isolation of the mid-Pacific and many already are envisioning a future in which their islands will be a junction point for fast passenger and freight services to Oceania, Australia, and the Far East."

The *Minneapolis Journal* wrote that the flight "...opened an aerial avenue over the eastern reaches of the Pacific, even as Lindbergh opened an aerial avenue over the Atlantic."

Lindbergh proclaimed, "It was one of the most perfectly organized, yet most difficult flights ever attempted."

Maitland and Hegenberger returned to San Francisco to be honored with a triumphant parade. Their successful flight grabbed some of the allure away from the Dole Race of course. But James Dole explained that since these men were military aviators, the Dole race remained a test for civilian aviation. Because of the prize money involved and the fact that the pilots of the Dole Race were "everyday people," the newspapers continued front page coverage of the pre-race preparations and the many foibles associated with the event. The public remained captivated by the prospect of a race of airplanes across the Pacific.

After his historic flight, Maitland continued his career in the Air Force, and in his mid-forties, flew forty-four combat missions in World War II for the 386th Bomb Group out of

Boxted Airfield in England. For his brave and exemplary service, he received the Distinguished Flying Cross. This was his second as he had received the Cross during World War I as well. In 1932, Hegenberger made the world's first all instrument flight at Wright Patterson Air Force Base, in Dayton. After WWII, Hegenberger continued with the Air Force and is credited with the development of radio directional beacons.

7

ERNIE SMITH JUMPS THE GUN

Tagging along with Maitland's attempt at the flight to Hawaii was civilian Ernie Smith. Smith had gotten together with sponsors to purchase a used mail plane, a Travel Air, manufactured in Wichita, named *City of Oakland,* which he rigged with a new engine. Smith did not want to wait for the Dole Race, valuing the notoriety his flight to Hawaii would bring more than the prize money. So he, and his navigator, took off from Oakland within minutes of Maitland's departure on June 28. However after less than an hour, *City of Oakland* turned back with a faulty altimeter. The press was highly critical of the Smith effort claiming he was attempting to upstage the Dole racers, who were not permitted to start until August 12, taking away some of that race's glamour. His failed effort appeared to be less than competent.

Ernie Smith and his financial backers were highly disappointed in not being able to continue on June 28. They had a faster plane than Maitland and had predicted they would be the first to land in Hawaii. When the *City of Oakland* was unable to continue, disputes among the backers including Smith, were played out in the newspapers as each sought to publicize their grievances against the others. But finally a majority agreed there would be another attempt. That decision was also controversial. One of the backers withdrew support and even threatened to sue to stop it. The other backers hired a new manager and Smith's original navigator, who resigned after the failure, was replaced by Emory Bronte. Bronte was renowned as the author of a manual on aerial navigation.

On July 14, 1927, shortly after 10:00 a.m., the *City of Oakland* rolled down the runway toward a takeoff. Maitland and Hegenberger had returned from Hawaii and were among the 10,000 spectators lining the Oakland airfield. The plane was loaded with enough fuel, according to calculations, for a flight of thirty-three hours. The extra weight hindered its handling. During the takeoff, after about two hundred yards the plane veered off the runway and came to a stop with one wheel stuck in a ditch. This was a less than a stellar beginning. An inspection showed no damage and the plane was wheeled back into position. At 10:39 a.m. the plane was finally airborne on its second attempt.

On this flight there were no ships stationed along the route, but ships in the vicinity of the route to Hawaii were notified to be on the lookout for Smith. The *City of Oakland* was equipped with a two-way radio. Radio operators in the area were advised to transmit Morse code messages no faster than ten words per minute because Smith and Bronte were not very proficient at operating the radio. As an interesting publicity stunt, the plane took several carrier pigeons ostensibly to carry messages from the middle of the Pacific Ocean to Hawaii.

After two hours the radio receiver went out and no more messages from below could be heard. They could send messages, but Smith could not be sure the radio was actually broadcasting. A carrier pigeon was released after two hundred miles and again after four hundred miles. Other than the radio, the plane operated well through the first day and night. Toward dawn however, the engine began running unevenly. Smith thought they were running low on fuel. They sent out radio messages. While they calculated there were at least six hours to go, the sputtering engine was revealing a shortage of gas. A SOS distress call was broadcast along with their location. Smith was sure he was going down. At one point Smith took the plane down to within a few feet of the water, thinking that it would be better to make a powered landing rather than wait until the engines stalled forcing him to glide the plane to the water. A powered landing would give him better control. Just then, the engine came back to life. He took the plane back up to 3,000 feet and continued on. Shortly after the SOS, the radio went out completely. Smith had cruised so close to the water, the wire antenna trailing beneath the plane had been torn off by the waves bellow, and now there was no transmission or reception from the radio. In the meantime, several ships responded to the SOS and steamed to the position indicated in Smith's call for help. The *SS Wilhelmina* reached the spot, but of course, found no sign of *City of Oakland* as Smith had gone on.

Smith continued toward Oahu, and as he came to the island of Molokai, only about thirty miles from his goal, his engine stopped completely. There was no question his fuel was now entirely gone. He released the last of the pigeons. None of the pigeons released during the flight made their way home. Smith made the decision not to land on Molokai's flat sandy beach. If the wheels dug in, the plane would likely flip. Instead he put down in a stand of six to ten-foot-tall kiawe trees, making a dead stick landing. Smith took the plane down in a slow glide. It approached in a shallow, silent trajectory. As they sailed into the stand of trees, all hell broke loose. The plane was bounced around and forcefully torn apart, ending up sideways in the brush. Neither Smith nor Bronte was hurt and they crawled out of the wreckage. Since no fuel remained, there was no fire. They had flown only about twenty-six hours, far short of the thirty-three hours expected. Smith could not account for their shortage of fuel.

Smith and Bronte were criticized for their lack of preparation and the fact they had sent an SOS and then continued on. However, it was not possible for them to broadcast a retraction as the antenna had become detached. They were criticized for attempting to overshadow the Dole Race. Their flight was disparaged because they did not make it to Honolulu and instead had crash landed. As a result, their accomplishment was discounted. But as author David H. Grover states in his article in *Aviation History* magazine, the flight of *City of Oakland* was much better prepared than the Rodgers flight of 1925. They were the first civilian aviators to actually make it all the way to the Hawaiian Islands. The fact that the flight ended in a crash contributed to the public's indifference toward the flight and the feeling that it had only limited success. As Grover points out this is probably unfair.

After his Hawaiian flight, Ernie Smith went on to become a pilot for TWA after flying in Howard Hughes' epic film, *Hell's Angels* which recreates aerial dogfights from WWI.

The Dole Race, scheduled to begin in only four weeks after the Ernie Smith flight, was a much bigger spectacle. Smith's flight was labeled as inept and soon forgotten. The public and the press turned its attention to the Dole Racers. They were interviewed and photographed on a daily basis. Anticipation grew as the Dole Race approached.

ꝋ

The Fliers Arrive

The Oakland Airfield was a work in progress. The City had just purchased the land in April. There were no permanent buildings, except for a small wooden shack housing a takeout sandwich counter. The runway was a recently excavated, long dirt strip that generated clouds of dust. Virtually no earthwork was complete other than the runway. From the air it resembled a huge funnel. The east end was over five hundred feet wide, providing room on both sides for planes being readied for takeoff. Just behind the east end was a huge swamp. As the strip proceeded to the west toward San Francisco Bay, it tapered to a thin strip just over fifty feet wide. The runway ended within two hundred feet of San Francisco Bay. The adjoining land with its shrubs and stunted trees was untouched.

Crews preparing for the race set up large, open-sided tents in which to work. Temporary wooden fences were erected around the perimeter to keep the milling spectators from overrunning the grounds.

Originally, not all the planes planned to take off from Oakland. Some pilots planned to leave from Crissy Field, attached to the Presidio (Spanish for fortress), an Army Base in San Francisco, which is ten miles closer to Hawaii. The Presidio closed in 1974, and is now a park overlooking the Golden Gate. Pedlar wanted to take off from Long Beach. But the Dole Race Committee mandated that all entries would depart from Oakland under the Committee's watchful eye.

On August 8, the low winged monoplane, *Humming Bird*, a Tremaine-Thalheld Pacific J-30, left San Diego for Oakland. The plane took off without incident and headed west toward the ocean. It was foggy and apparently the pilot lost his bearings. Shortly after takeoff, *Humming Bird* crashed into the embankment along the shore at Point Loma and burst into flames. Navy Lieutenants George Covell and Richard Waggener, the pilot and co-pilot, were killed. Some reports indicate the name of their plane was to be changed for the race to The Spirit of John Rodgers. Coincidentally, *Humming Bird* had won starting position thirteen in the drawing held that very day.

By Tuesday, August 9, three days before the starting date, only eight planes had arrived at Oakland. The others were either "on their way" or considering withdrawal from the race due to financial and other issues. All these details were reported in newspaper stories across the United States. On that date, the banner headline in the *San Diego Union* newspaper was, "Five San Diegans Enter Dole Flight." The paper named Kenneth Hawkins and Norm Goddard, with *El Encanto*; George Covell and Leo Pawlikowski with an unnamed entry; and Manley Lawing with *Miss Doran*. All the men were San Diegans from the North Island Naval Air Station. A lower, smaller, headline announced that President Coolidge was not seeking re-election as President. Apparently the editors considered that the Dole Race was the more attention-grabbing story and deserved the lead headline.

On Wednesday, August 10, the Race Committee in Oakland called a dinner meeting to consider whether to postpone the race. It was only two days before the race and not all the entrants had arrived. The earlier arrivals had not yet been approved. It would be impossible to inspect and test the planes and their crews arriving on Thursday for the Friday race.

Augie Pedlar threatened to start his race on the 12th regardless of the Race Committee decision. "They can keep the prize money. Of course, I would hate to see some of the planes take off in their present condition, but everyone had time to prepare and what is fair for one is fair for all. We are ready and we may go tomorrow regardless of last minute decisions by the committee."

Malloska chimed in from Hawaii, "We have spent thousands of dollars in preparation for this flight and are ready to go. To postpone it would be unfair. Pedlar will fly anyway."

The Race Committee in Honolulu was in charge of the welcoming ceremonies for the successful racers. The dignitaries planned the official welcome for Saturday. They wired the Oakland Committee. "The Race Committee [in Honolulu] disapproves postponement of zero hour for starting time. The contestants have already been given sufficient time to prepare." The Race committee withdrew its proposal to postpone the race for two weeks. Instead, it proposed a four day delay until the following Tuesday. Despite instructions from Honolulu, shortly after midnight on Thursday, August 11, the Oakland Race Committee announced that the race was postponed until Tuesday, August 16. Honolulu would have to wait until Wednesday.

Earlier in the afternoon on Thursday, August 11, the plane sponsored by western film star, Hoot Gibson, crashed into the bay during pre-race testing. The International Triplane powered by two engines, the *Spirit of Los Angeles*, had been nicknamed, "a stack of wheats" because of its three wings. All three of the crewmembers aboard were rescued, but the plane was damaged beyond repair. Hoot Gibson was furious. The Oakland Race Committee became even more alarmed due to the crashes and redoubled its efforts to inspect the planes for safety.

In light of the decision to postpone the race for only four days, almost all of the pilots signed a "gentleman's agreement" to start the race on August 16 rather than the 12th. Only Frank Clark, pilot of *Miss Hollydale,* an orange and black biplane, refused to sign. Clark was a movie stunt pilot and wing walker. On Saturday morning, August 13, without any prior warning, Clark and a financial supporter, Charlie Babb, took off and were last seen heading west through the Golden Gate. Rumors flew around the field. What was Clark up to? There was speculation that since he obviously disagreed with the postponement, he was getting a head start to Hawaii. However, the following day Babb sent a wire to the Race Committee explaining that he and Clark had shuffled home to Los Angeles and were withdrawing from the race.

Miss Doran, after being repaired in the Mendota wheat field, made her way to Oakland. When the *Miss Doran* arrived on August 6, last minute modifications and fuel calculations were made. Manley Lawing, *Miss Doran's* appointed navigator, was an experienced Naval navigator. He lacked experience in aerial navigation, however. He went up to demonstrate his skills to the Department of Air Commerce and reportedly got lost flying over the City of Oakland. Clarence Young ordered Pedlar to find a new navigator. Vilas "Cy" Knope was recommended. Lieutenant Knope, a graduate of the US Naval Academy, had a reputation as an accomplished Naval aerial navigator. Pedlar contacted Knope in San Diego and asked him to come to Oakland. Pedlar was not as prepared as he boasted two days earlier.

Finding a navigator was not the only problem facing Pedlar in the days before the race. On another pre-race test flight the earth inductor compass on *Miss Doran* was found to be off by seventy five degrees and had to be recalibrated upon landing.

As the start of the race neared, reporters in San Francisco wrote story after story about the fliers. Their favorite subject was Mildred Doran. One story started with the headline, "Dole Hop Girl Composed As Race Nears" and continued on the back page with the headline, "Girl Flyer In Dole Race, Cynosure [the center of attention] Of All Eyes." The article is almost poetic. "Mildred Doran was dressed in a smart tan sports coat and close fitting hat, barely distinguishable from a hundred other slim young girls on the field. But one look into her brown eyes and you knew she was booted and spurred for the Out Trail, the trail that is always new...." When asked if she would be returning to the classroom, her reply was, yes...well maybe. Linnets [a species of songbird] are not made for the classroom. They fly as sparks, they go upward." One cannot help but realize that she did not think through the metaphor, as sparks typically vanish as they fly toward the sky.

Mildred opened the mica glass door to display her post in the plane. She explained that she would sit behind the pilot, able to communicate with Pedlar and Knope by speaking through a megaphone. There were gas tanks between them. A makeshift toilet was hidden under her seat. She smiled as she explained, "people are always curious about that." The seat was covered with an inflatable cushion that could double as a flotation device. Her compartment was small, but so was she, and so, she had plenty of room.

As the race approached, Mildred expressed fatigue from the pre-race activities. She was tired of being "stared at" and being asked "hundreds of questions." She had lost five pounds in the last week and remarked, "I have faced the lens so often I feel as if I am posing in my sleep."

Mildred was not the only female planning to be a part of the race. The spouse of William "Lonestar Bill" Erwin, Connie Erwin, planned to fly with her husband in *Dallas Spirit*. But, the race committee disqualified her two days before the August 12 start date due to her age of nineteen. The Committee decided that all entrants had to be at least twenty one. Other reports mentioned another reason. She was pregnant. Pauline Rich, another woman declared early on her intent to fly in the race with husband, Garland Lincoln, but their sponsor was not able to purchase a suitable plane.

On August 12, British pilot, Frederick Giles, announced he would leave Detroit to join the race. He attempted to take off, but could not get his Hess Bluebird, *Detroit's Goodwill Messenger*, off the ground. He later announced a more ambitious plan. In November, Giles took off for New Zealand from Mills Field in San Francisco, with a scheduled stop in Hawaii. He was forced to turn back after five hundred miles, after encountering a severe storm causing a tail spin. In the spin he spilled his instruments, charts and food into the ocean. He made it back to San Simeon, California, where he landed the plane. Giles was unhurt. His claim that a "storm" caused his return was disbelieved by agents in the San Francisco Weather Bureau.

On Saturday, August 13, during a test flight in Los Angeles, the entry of Arthur V. Rogers, *The Angel of Los Angeles*, unexpectedly went into a dive as he flew at low altitude directly over a runway filled with spectators. The crowd gazed up in dismay. Rogers, a WWI ace, scrambled and pulled himself out of the open cockpit. He jumped clear of the plane wearing a parachute. Unfortunately it failed to open. To the astonishment and shock of the spectators and his family below, he fell one-hundred-and-fifty feet to his death, landing close to where his plane had nose-dived into the runway.

The race had not yet begun, and there were three crashes and three fatalities. But, it was never reported that any of the remaining racers considered withdrawing based on these accidents. It's possible that Clark took *Miss Hollydale* home due to safety concerns, but he never publicly disclosed his reasons.

On Saturday, August 14, two days before the race, Pedlar went up with CY Knope and a Department of Commerce representative. Pedlar and the Department were pleased with Knope's abilities as a navigator and Pedlar hired him as the new navigator of *Miss Doran*.

Mildred Doran came to the field on Sunday afternoon upon leaving church. Even though the race was two days away, there were thousands of spectators out to get a glimpse of the flyers and their aircraft. Reporters rushed to speak to Mildred. The crashes of the last few days were on their minds. When the press asked her if she was afraid to make the daunting flight to Honolulu, she replied, "Life is full of danger." She was sure she would be the first "girl" to achieve aeronautical history. She continued, "If I wasn't certain we could make it, I would not be foolish enough to start out." As she demonstrated to reporters her assigned station for the flight, a small khaki colored cushion, and the small hole which leads to the pilot's quarters, she lamented, "Oh, how I want to get started. How I want to get started."

A newspaper printed a 'human interest' story about the race: "Buzz-z-zz went the speaking tube in the kitchen. The boss chef wiped his hands on his apron and answered. 'Lady in 496 wants twelve chicken sandwiches and three quarts of coffee,' a voice said. 'My golly,' exclaimed the chef. 'For one lady?' Why don't she join the circus.' 'Send 'em up Tuesday morning' the chef was told. 'They're for Miss Mildred Doran, the Honolulu flier.' In this way, the Flint schoolma'am disposed of her 'detail' in the Dole race. Being a woman, she had been given the task of putting up the lunch while pilot Augie Pedlar and V.R. Knope, the new navigator made other preparations."

On the afternoon of the 15th, the Gideon Society handed out bibles to the crews as they gathered around their aircraft. Augie Pedlar spoke to reporters and denied a rumor that Mildred was considering withdrawing from *Miss Doran's* flight.

Reporter Ethel Bogardus described the scene on the night preceding the race: "Brooding quiet disturbed now and then by the blare of an automobile horn, darkness settled over the field." Nearer floods of light poured out a little way into the blackness, illuminating the blunt noses of great planes, where men in greasy coveralls worked silently, except for brief commands given. The light shone in half tones on the wings throwing the fuselage in shadow.

There they were in long, uneven lines, waiting, with the great planes resting before their long flight. A wooden fence guarded them from the crowd that hung over the protecting railing, watching quietly , eagerly, while the great birds were given their last preening. Officers in uniform, their stars gleaming through the dusk, paced slowly about their precious charges. Wrapped in her nightcap, the *"Miss Doran"* waited. Ready. No mechanics needed for last minute touches — a stubby ship, with a humped back. Behind that white door in the fuselage, a girl was to ride to Honolulu.

A chug-chug sounded suddenly out of the darkness. A tractor dragging over the field, smoothing the long runway was doing its last bit for those who would speed down it in a few short hours.

Across the field at the top of the runway a wide chalk line ran straight. And over there the Oklahoma spreading dark wings was ready to go. It had taken its place on the starting line, its nose headed into darkness.

Weary men watching. Great planes poised ready to fly. Waiting for whatever fate the morrow might bring."

Throughout the night before the race, work continued under floodlights on three of the entries. *Woolaroc* was loaded with gasoline just arrived from the Phillips refinery in Bartlesville, Oklahoma. Marty Jensen, planning to rely partly on five gallon cans of gas to keep his plane airborne ordered a new gasoline pump installed on *Aloha*. Mechanics worked on it well past midnight. Jimmie Irving had his plane, *Pabco Pacific Flyer*, loaded with food supplies during the night, being careful to store the victuals in the order they would be needed.

A Gallery

Of

Pictures

Richard DuRose

Mildred Doran

Richard DuRose

Marty Jensen in 1929

Mildred Trying On A flying Cap

Richard DuRose

Mildred and Honolulu

Mildred On The Wing of *Miss Doran*

Vilas "Cy" Knope, Mildred Doran, and John "Augie" Pedlar at Oakland Field

(L to R) Bill Malloska, John "Augie" Pedlar, Manley Lawing, and Mildred Doran with Miss Doran

Richard DuRose

OFFICIAL PROGRAM

DOLE
TRANS-PACIFIC
FLIGHT

North America to Hawaii
August 16, 1927

Program under the Auspices of
OAKLAND CHAPTER
NATIONAL AERONAUTIC ASSOCIATION
Oakland, California

PRICE - - TEN CENTS

BENNETT & MOREHOUSE   359 19TH ST., OAKLAND

Souvenir Program from the Dole Race

Mildred On A Jumper

Richard DuRose

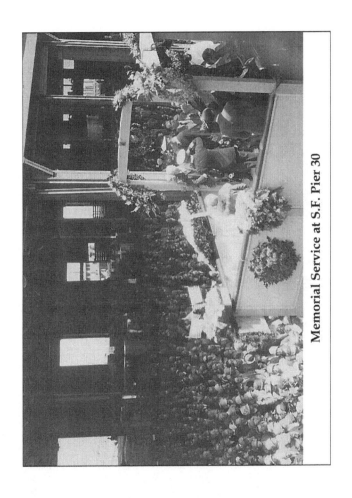

Memorial Service at S.F. Pier 30

Souvenir given to those who flew with Lincoln Standard pilots

Richard DuRose

Ceremony aboard SS Maui

Flower arrangement from Mildred's
students at Caro Elementary

**Mildred Standing In Front
of Miss Doran**

Richard DuRose

9

They're Off!

On the morning of Tuesday, August 16, the day of the race, a steady ribbon of cars streamed into the airport. By noon, an estimated 50,000 folks engulfed Oakland Field. Many had skipped out on their jobs producing a holiday atmosphere. The onlookers brought picnic lunches, binoculars, and portable stools. The National Guard and police kept the throngs of onlookers off the runway and behind the fences. More spectators stood on rooftops or on boats bobbing on the bay.

The crowd buzzed with the news that one of the entries, the Air King biplane *City of Peoria,* and its pilot, Charles Parkhurst, had been disqualified that morning. The biplane had been brought to the starting line at 9:00 am, but after its fuel capacity was studied, it was disqualified by C. W. Saunders, the head of the starting committee. This left eight planes prepared to begin the race at noon.

Several ambulances were stationed along the runway. A thirty man "crash detail" from the 381st Service Squadron of the Air Corps Reserves from San Francisco assembled at the starting line prepared to wheel the planes into starting position. And, if there were any mishaps, they would assist in rescue operations, fighting fires, and rescuing downed fliers from the San Francisco Bay.

Mildred left her hotel, making sure to thank the staff for all their courtesies. A reporter commented that she appeared to be a favorite guest. As she arrived at the airport, she was surrounded by spectators. She answered the innumerable questions with grace and patience. She shook hands, autographed postcards, and gave a short talk over radio station KLX. She claimed the radio interview made her more nervous than the flight. Before stepping on board *Miss Doran*, she placed her father's last minute telegram of encouragement in her jacket pocket.

Augie Pedlar wore his trademark, checkered knickers, a mismatched plaid jacket, and a well-worn straw skimmer, sometimes called a "Kelly." Pedlar, at twenty-four, was the youngest pilot in the race, but due to his limp, pale complexion, and balding hair, gave a much older appearance. He had a ready smile which displayed a slight gap in his front teeth. Cy Knope, sporting a thick mustache, was sharply dressed in his naval officer's uniform. Mildred appeared in a tailored, military looking, flying suit with khaki jodhpurs, trousers fitted tightly around the lower leg, and puttees, cloth wrapping below the knee. The tunic, complete with Sam

Browne belt, was embroidered with Lincoln Oil insignias on the sleeves. The left pocket sported several fraternity pins. Her outfit had been designed and made by a tailor from Michigan. The three stood next to the *Miss Doran and* posed for last minute pictures. They were all smiles. Mildred followed the men into the plane, turning to wave at the crowd just before ducking through the door.

On the morning of the race, the usual sea breeze was missing. Early on, a thick fog had rolled in from the Bay, hovering about three hundred feet above the runway. Planes that went up for one last practice run quickly disappeared into the mist. Because of the extra weight of the fuel, the aircraft were unsteady and difficult to maneuver. A wind, however slight aids in takeoff by creating more lift. The lack of wind was going to be a problem.

Tractors paraded up and down the runway administering a last minute rolling to maintain a smooth surface. Dust bellowed up from the far end of the runway. Water trucks followed, spraying in an attempt to keep the dust under control. Motorcycle police escorted the arriving crews to their planes.

The original starting order had been changed slightly to allow the planes that had qualified first to take off first. The crash team assembled the entrants in order at the east end of the runway. Shortly before noon, eight thunderous aircraft engines roared to life. Ambulances and fire trucks found their spots about half way down the runway.

At precisely noon, Ernie Smith, the pilot that had made it to Molokai a month earlier, stood out in the runway with a large checkered flag, which he dropped from above his head in dramatic fashion. Fifty thousand onlookers crowded closer, straining to see over or through the fence. As the flag came down, the crowd roared so loud it drowned out the loud howl of the eight aircraft engines. The race was on.

The Travel Air, *Oklahoma,* had been wheeled into place. Within a half second of Smith's dramatic gesture, *Oklahoma* started down the runway, pushed for the first few feet by its handlers initially at a walk and then running until they could no longer keep up. As the plane rolled down the runway picking up speed, the crowd strained to see. Many climbed up on their running boards to get a better view. Many waved. At 12:02, the *Oklahoma,* with pilot, Bennett "Benny" Griffin, lifted off and headed west over San Francisco Bay. The crowd of 50,000 let loose a loud spontaneous cheer, as it would with each successful takeoff. Next, the silver *El Encanto,* [the charm] with its designer, Norm Goddard at the controls started down the runway with *Oklahoma* still in sight. It slowly gathered speed. Once or twice it seemed about to rise. At about the half-way point, the plane made a slight right turn followed by a left, and then a more violent right turn again. It made a ground loop with one wing touching the ground causing a sudden rotation. It ended up in a hollow about one hundred feet off the runway. The crowd let out a collective "Ooooh!" The *El Encanto* ended up on its side in a heap by the side of the runway with one wing pointing toward the sky and the other wing crumpled beneath her. Goddard and his navigator crawled out, but the plane was ruined. *El Encanto* remained a broken mess, its nose pointing back to the starting line, for the rest of the day. Goddard sheepishly told reporters, "I'd rather have gone down in the ocean than to have this happen."

Next to start, at 12:05 was Livingston "Jimmie" Irving, World War I air ace, in the Pabco *Pacific Flyer*. The plane was a Breese high-wing monoplane painted bright orange with black trim. Irving had been a pilot in WWI with the Lafayette Escadrille, a squadron of the French Flying Corps made up mostly of American volunteers. The plane was sponsored by the Paraffine Companies, manufactures of fiberboard, using the trade name Pabco. As Irving moved down the runway, it was obvious that the plane was struggling under the weight of the hundreds of gallons of fuel it was carrying. One observer said, "It waddled like a duck." Just as the plane moved out of sight of the crowd, the plane leaned to the right. Acting fast, Irving shut down the engine and coasted to a halt. The plane sat silent and unmoving on the runway. The spectators strained their eyes to see the plane airborne, but nothing appeared. Rumors of its fate circulated. After twenty minutes, a truck was dispatched, and the Pabco *Pacific Flyer*, was unceremoniously towed back along the edge of the runway toward the starting line.

Jack Frost was forced to wait while *Pabco Pacific Flyer* sat in the runway. Finally, his *Golden Eagle*, a gold and blue Lockheed Vega, set off down the runway rolling past the *Pabco Pacific Flyer* being towed back. At 12:31, its wheels left the runway after about 4,500 feet to another loud cheer. *Golden Eagle* had easily lifted off and quickly made altitude. The *Golden Eagle* was owned by the William Randolph Hearst family, wealthy newspaper publishers.

Next to be wheeled into place was the *Miss Doran*. As she rolled down the runway, another cheer echoed across the expanse of the airfield. The silhouette of Mildred could be seen in the rear window. At 12:32, *Miss Doran* with four-hundred-and-seven gallons of fuel, and thirty pounds of oil, was airborne leaving the runway at about 3,000 feet. Everyone watched as she silently faded out of sight over the Bay.

At 12:34, Hawaiian pilot, Marty Jensen, in the bright yellow, *Aloha,* took off in a shallow trajectory. As the plane reached the very end of the runway, it was barely airborne, and its wheels brushed across the sandy terrain between the runway and the Bay. As she started out over the Bay, *Aloha* remained low to the water. At 12:35, Art Goebel in *Woolaroc* took off and began its upward glide. In an act of showmanship, Goebel took *Woolaroc* over San Francisco's Market Street at 1,500 feet before heading out over the Pacific. Finally, "Lonestar" Bill Irwin's green and silver, Swallow, named the *Dallas Spirit* rolled out and took off successfully at 12:37 after rolling about 3,500 feet from the starting line. With him was his newly recruited navigator, Ivan Eichwaldt.

The Race Committee sent a two-word cable to Honolulu: "They're Off!" The big siren in the Aloha Tower on Honolulu's waterfront let out a high pitched screech. Because of the excitement, little business would be transpired on the island until the fliers landed. The people of Honolulu prepared for a sleepless night waiting for the fliers to arrive.

Oklahoma, Golden Eagle, Miss Doran, Aloha, Woolaroc, and *Dallas Spirit* were on their way. *El Encanto* was wrecked and would not join the race. *Pabco Pacific Flyer* was being worked on by its crew. The crowd began to disperse. The spectators began to wander away from the fences and back toward their automobiles.

Then, there was a murmur followed by some excited shouts. A plane was seen in the distance. At first no one could make out who it was. Then, it became clear—it was the biplane. The *Miss Doran* was coming back downwind toward the field with gasoline streaming from beneath the fuselage. Augie Pedlar wanted to lighten the load before landing to improve his maneuverability. The plane came in and taxied straight to her mechanics' station. Cy Knope jumped out and declared that the engine was backfiring like a "tin lizzie." Tin Lizzie is the nickname for the Model A Ford, notorious for loud booms caused by exploding fumes in the exhaust. Mildred, struggling to keep her emotions in check, was escorted by Ernie Smith to a nearby tent to avoid the crowd. Ernie patted her on the back and told her, "Don't let this stump you, Mildred. Get the old boat fixed up and get going again. The same thing happened to me, you know, and I got there." As they made their way to the nearby tent, Smith admonished the surging crowd, "Leave her alone fellahs, leave her alone."

As the mechanics began to work on the *Miss Doran* another shout went up. One more plane was spotted by the crowd making a bee-line toward the airfield. It was the *Dallas Spirit*. As it approached, it became obvious why Bill Irwin was returning. A huge rip in the canvas fabric covering the right side of the fuselage was flapping in the wind. Recent modifications had surreptitiously started a tear in the fabric. *Dallas Spirit* landed and quickly taxied to its waiting team for damage assessment.

At this point, the crowd spotted yet another dot on the horizon. The crowd was astonished to see so many planes returning to Oakland Field. How could it be that a third aircraft was quitting the race this early? It was *Oklahoma* trailing a thick trail of black smoke. It had five blown cylinders. *Dallas Spirit* and *Oklahoma* were through, for that day at least.

At 1:22 pm., Jimmie Irving made another attempt to take off in the *Pabco Pacific Flyer*. Half way down the runway, the plane went airborne by almost twenty feet. Its flight was short lived, however. It came back to the runway—hard. The crowd let loose a spontaneous scream. One wheel touched down. The plane bounced up, then fell back to the runway. This time it went into a long ground loop, and ended up on its side with most of its wood and canvas tail in shreds. Its bottom was torn away. Irving was unhurt, and when his wife was rushed to him at the end of the runway, there were tears in his eyes. He looked up and said, "Well, my dear, I won't get to Honolulu now."

Repairs were made to the *Miss Doran* amid some confusion. A photo taken from the air while the *Miss Doran* was undergoing repairs shows about fifteen men surrounding the front of the plane near the engine, two trucks backed up to the left side of the plane, and a forty-foot diameter circle of about three hundred spectators surrounding the plane. The crowd was noisy and the mechanics were working at a frenetic pace to get the plane back into the race.

Frances Russell of the *Oakland Tribune* reached Mildred in the tent and asked her why they had returned. She replied, "I don't know why we came down, but I know we are going back up again. You have to take things as they come. There's no use making a fuss about it." Wishing to get out in the air, she put on some clothes given to her by friends so she would not be recognized and hassled during her wait to re-enter the race. Mildred promised to take the friends up for a spin when she got back. "It's nothing more than riding in a fast elevator."

New spark plugs were installed in the plane's engine. Some tried to convince Augie Pedlar to give up as it was not clear that spark plugs alone would solve the problem. Only a few days earlier he had changed spark plugs after the emergency landing in Mendota, California. Augie was red-faced, agitated, and determined. As recounted in the book, *Glory Gamblers*, by Leslie Forden, a man who was there said he could not recall that anyone topped off the gas tanks before *Miss Doran's* second try. Cy Knope went to the tent to fetch Mildred, and told her that he and Augie thought it would be a good idea if she did not go. Mildred reacted — if the plane was going, so was she. Some reported seeing her in tears as she re-boarded the plane. Others thought her face appeared ashen. Speculation that she was fearful is at odds with her statements to Frances Russell mentioned above. Finally, the repairs were completed and it was time to try again. The *New York Times* described the moment as follows:

"The ardor of the race was gleaming in Miss Doran's hazel eyes when she climbed into her red, white, and blue biplane for the second time in the afternoon and bade her pilot streak through the lonely air lanes over the ocean and catch Jensen, Frost, and Goebel if they could."

Just after 2:00, ninety minutes after her first attempt, the *Miss Doran* rolled out to the runway amid the loudest cheer of the day, and took off with a thunderous growl. The crowd watched as the *Miss Doran* flew off over the bay heading west. Soon, it was out of sight. At 3:03 pm, *Miss Doran* was last spotted passing over the Farallon Islands, twenty seven miles off the coast.

The National Aeronautics Association notified Honolulu by cable that four out of the eight planes were on their way. The crowd waited for an hour to see if *Oklahoma* and *Dallas Spirit* would try again, but finally drifted away. Oakland Field was quiet.

As an aid to the racers, the Matson steamship company issued instructions to two of its liners, SS Manukai and SS Manulani to flash their searchlights every ten minutes throughout the night.

Various sightings of the planes were reported by ships stationed along the route. 2:30 p.m.: Motor ship *Silver Fir* reports spotting *Aloha* 185 miles out, north of course. 2:43 p.m.: *Miss Doran* spotted flying over the Farallones. 2:50 p.m.: *SS Wilhelmina* reports seeing *Aloha* heading south toward a more direct course. 2:55 p.m.: Destroyer *Meyer* reports seeing *Aloha* 200 miles out still north of direct course. 4:00 p.m.: Destroyer

Hazelwood spots *Woolaroc* 270 miles out. 4:35 p.m.: *SS Wilhelmina* reports a radio message from *Woolaroc* 300 miles out. 8:00 p.m.: Destroyer *McDonough* reports *Woolaroc* 390 miles out, "Seven hours out all going fine." 8:50 p.m.: Destroyer *Cotty* reports radio message for *Woolaroc* 750 miles out. 2:00 a.m.: *SS Manulani* reports two planes believed to be *Woolaroc* and *Miss Doran* to be about half way (1,200 miles) on the northern edge of the course. 2:00 a.m.: *SS City of Los Angeles* reports messages received by Army signal corps from *Golden Eagle* and *Aloha* on the southern edge of the course. 4:00 a.m.: *SS Manulani* reports *Woolaroc* 1,485 miles on course. 8:59 a.m.: *City of Los Angeles* reports *Woolaroc* 459 miles from Honolulu. 9:30 a.m.: *City of Los Angeles* reports *Woolaroc* 450 miles from Honolulu. The accuracy of these reports may be questionable.

Hawaiians were excited by the Dole Race. *The World*, a New York newspaper, described the panorama at Wheeler Field, an Army Air Base on Wednesday, August 17. The headlines read: "Honolulu, Gone Race Mad, Flocks to See Finish. Crowds Begin to Assemble at Wheeler Field Soon After Midnight..." A description of the scene followed: "The Pacific's greatest aerial rodeo brought out all the color and emotion that have made these tropical islands famous. Long before the sun had peaked over Diamond Head this morning, a stream of humanity had found its way to Wheeler Field to witness the arrival of the planes in the James D. Dole $35,000 race from Oakland.

The crowd was noisy, but good natured. All races in the territory were represented. Pretty little Japanese women in sashed kimonos, Chinese in jackets and pajama pants, Polynesians and Coreans (sic), Filipinos, and Nordic blondes made up the welcoming multitude.

Parking space had been provided by the army officials for 8,000 automobiles. By 8:30 a.m. . . . . this space was filled and the Territorial Highway was lined with additional cars for a distance of two miles. Army estimates of the crowd grew hourly. At noon, it was stated by the Reception Committee that more than 20,000 people had crowded into and around the field....

The day was preceded by a hectic night. Theatres [were] kept open to amuse those who could not go to sleep and to keep them informed of the progress of the flyers. Bulletins were announced in the theatres.

The reviewing stand was early filled with notables, headed by Gov. Wallace R. Farrington, and high army and navy officials. Many scout and patrol planes skirted the field at times doing stunts and watching for the racing planes. A band played popular airs. Field kitchens had been placed in strategic positions and they were doing a land office business."

Richard DuRose

10
VICTORY FOR GOEBEL

Art Goebel in *Woolaroc* was the last to take off, other than *Miss Doran's* second attempt. Goebel was born in a small town forty-five miles from Albuquerque, New Mexico. At an early age he developed an interest in mechanical things, starting with his bicycle, which he took apart and put back together. As he grew older he bought motorcycles and later, automobiles. He took apart each new vehicle and put it back together to determine how it worked. He became a talented mechanic. He planned to leave his small town life and see the world. When Goebel was fifteen year's old, a pilot brought a plane to Pueblo, Colorado, where Goebel lived at the time. Goebel was enthralled with the aerial demonstration, although he was not immediately drawn to an aerial career. That would come later.

In his late teens, Goebel began trading for automobile parts. After about a year, he had accumulated parts from many different types of automobiles. He then took those parts and built an automobile that ran. In the process he learned about the inner workings of the internal combustion engine. As he says in his autobiography, *Art Goebel's Own Story*, once he had an automobile, "...New distances opened up to me, and of course, new comforts as compared to the motorcycle."

In 1918, Goebel joined the Army and went to France during World War I. He was a "good shot" having spent time hunting as a boy, and was assigned as a rifle instructor. After the war, Goebel thought, "I can't go back to the farm." He worked in Fort Worth and then Denver as an auto mechanic for Cadillac.

Goebel explains what happened next. "The war, and aviation exploits overseas as well as at home, had put air-talk on everyone's lips. Somewhere, somehow, I had known I was going to learn to fly—in recent years there had been no doubt about that—but when the right opportunity would come, I had not known." He quit his job in Denver and went to Los Angeles where he took a job as an aviation mechanic. Then in 1922, he went to work for Douglas Aircraft Company on the assembly line.

In his spare time, Goebel learned to fly. But that was not enough. He wanted to own a plane. As he says, thanks to the Army, he was able to purchase one, and then a second, and then two more "Jennies" which the government sold at low prices as surplus from the war. As he had done in the past, Goebel took apart and reassembled the mechanical parts of his

Jennies. He had a reputation for being fastidious about repairs and maintenance. His aircraft always operated without incident.

Goebel started thinking about how to make a living by flying, and took a job with an aerial photography outfit. Various businesses and individuals would pay to have their buildings photographed from the air. At that time, movie producers were hunting for the next big thrill and began using aircraft in film. Goebel became a stunt pilot for the film industry and gathered a reputation, primarily for his ability to fly upside down. He was adept at a stunt done by others as well in which the plane flies low over a moving automobile and rescues the hero or heroine by lowering a ladder from the plane to the car. Then, he invented a new stunt. The hero would be tied up and perched on the wing of a plane in flight. Goebel would swoop down, and with a hook, pull the hero off the wing to safety. Of course, the slightest mechanical issue in the midst of this stunt would be disastrous. Goebel prided himself in always having his equipment in good shape and running smoothly. These stunts called for precise and well-honed flying skills as well. It is interesting that Goebel, who gives every impression of being meticulous and careful, was willing to "push the envelope" and risk his life for the sake of movie stunts.

Goebel's next job was as a test pilot for Douglas Aircraft. He tested planes by putting the plane in a tail spin to "see how many times I had to turn it over to get it out." This was another job in which Goebel flirted with disaster.

When Goebel heard about the Dole Race to Hawaii, he contacted Frank Phillips of Phillips Petroleum in Bartlesville, Oklahoma. Phillips later used the brand name "Phillips 66". Frank and his brother, L. E. Phillips were interested in promoting their new lighter, more efficient aviation fuel and agreed to sponsor an entry for Goebel in the Dole Race. A new airplane was ordered from Walter Beech's Travelaire plant in Wichita, Kansas. Goebel went to Wichita to inspect the plane during assembly. When the finished airplane was delivered, it was named *Woolaroc* by Philips to celebrate the "woods, lakes, and rocks," of Oklahoma. Curiously, one newspaper article from that time period claimed *Woolaroc* is an old Oklahoma Indian word meaning "Good Luck."

Goebel arrived at Oakland only five days prior to August 12, the original start date. He was thankful for the four day delay. Ever the perfectionist, Goebel had Army experts build a two-way radio for the flight. The Army had recently begun to broadcast radio beacons from San Francisco and Maui. Goebel made several flights to test his ability to use those beacons as direction finders. The War Department had warned against reliance on the radio signals as they were considered undependable. *Woolaroc* was the only plane in the race with a working two way radio. *Woolaroc* with four-hundred-and-seventeen gallons of gas, seventeen gallons of oil, and plenty of extra equipment weighed 5,520 pounds on takeoff. Phillips Petroleum delivered fuel for Goebel from its Bartlesville refinery in newly designed refueling trucks.

Goebel's recollection of the flight was one of loneliness, especially once the sun set. He was separated from his navigator, Bill Davis by ten feet, and the space was filled with gasoline tanks, and they could not talk to one another. Instead they passed written notes now and then. There were clouds above and fog below. The almost full moon was not visible. The hours crept by. Every few hours, Goebel would take the plane up above the clouds so Bill Davis could verify their position. Fuel was running low. The radio began to act up. The engine droned on; finally, the islands came into view. Maui's mountains stood like a signpost for *Woolaroc*. Goebel and his navigator, Bill Davis let out a whoop! They were close enough now to make it on a glide even if their fuel did run out.

Army planes came up to meet them and one came alongside. The pilot held up one finger, and Goebel knew he was the first to arrive. After circling the field six times, they landed safely at Honolulu's Wheeler Field after twenty six hours, seventeen minutes and thirty three seconds in the air. They had won the Dole Derby.

Goebel, dressed in a rumpled suit, and Bill Davis, his navigator, wearing a blue Navy uniform, stepped out of *Woolaroc*, on shaky legs. They were quickly surrounded by the 20,000 boisterous well-wishers who had gathered at Wheeler Field, including the Governor of the Territory, Wallace Rider Farrington. James Dole announced that *Wooloroc* was the sanctioned winner. As Goebel and Davis walked up to him, Dole remarked, "I'm mighty happy, boys, that you made it safely." Governor Farrington told the crowd, "There is nothing I can say except that it is all very wonderful..." Hawaiian girls placed flower leis around their necks. Strident and enthusiastic marching music filled the air and the crowd cheered.

11

JENSEN JUST MAKES IT

Martin "Marty" Jensen did not know whether he would have a plane to fly in the Dole Derby until five days prior to the original start date of the race. Jensen was a native of the Hawaiian Territory and a stunt pilot. Even before Dole's announcement, Jensen had dreamed of making the flight from the mainland to Honolulu. Once the Dole Race was on, Jensen was determined that Hawaii would have an entry. He began to solicit funds from his friends using their pride in the Islands as the hook. Contributions flowed for a short time, but then dried up. He had enough—barely—to buy a plane and took a steamship to California, leaving his wife with a power of attorney. He almost did not have a plane to fly. He arrived at the Breese factory in San Francisco less than an hour prior to another prospective racer to gain first rights to a plane on the factory floor.

His Breese monoplane, *Aloha*, was wheeled out of the factory in San Francisco on Tuesday, August 9. Its bright yellow paint was barely dry and Jensen was proud of the Territorial Seal painted in blue on either side, as well as the colorful flower lei painted around its nose.

Aloha made its maiden flight on Wednesday, flying to Oakland Field. It was just two days before the scheduled August 12 race date. It was only after arriving at California that Jensen recruited a navigator for the flight, German born Paul Schluter. The *Aloha*, operating on a shoe-string budget, had no radio. Schluter was going to navigate by shooting the stars.

Aloha's wing and tail gas tanks were inadequate for the trip, so Jensen planned to carry an additional two-hundred-and-fifty gallons in the plane with him in fifty cans holding five gallons each. He planned to re-fill the main tank by hand during the flight. Jensen was ordered by the race committee to make better arrangements. Jensen could not afford to pay his mechanics who installed an extra fuel tank. They agreed to do the work on Jensen's word. Even after his new gas tank was fabricated and installed, he needed to carry several additional unattached five-gallon cans. On the morning of the race, Jensen received a wire for $300 with a note from his wife, Marguerite, saying there would be no more money coming. That was all she could raise. The mechanics were paid the whole $300. Jensen was left with a five-dollar bill in his pocket. Hearing about Jensen's predicament, Goebel and some others chipped in $25 so Jensen and Schluter would have enough to get a hotel and a meal in Honolulu when they arrived.

Jensen took off using the entire 7,070 feet of runway (1.3 miles) and barely lifted off as his wheels hit the sandy soil near the water's edge. He held the plane low and only gradually brought Aloha up. When he got to 100 feet he held her steady. Jensen and his navigator Paul Schluter had agreed before the flight to keep the plane below the fog. At one hour and twenty-eight minutes, while still flying low, Jensen flew directly over a cargo ship, pulling up at the last second, barely clearing the ship's masts. As night fell Jensen brought the plane up briefly to find a spot above the clouds for his navigator to be able to see the stars.

Night flying presents a particular danger of disorientation in which the inner ear fools the pilot into thinking that he is flying level to the horizon when he is not. Without any visual reference to the horizon, as the plane slowly tilts to one side, it begins a turn. Over time the tilt will become more pronounced. Only after the turn is accentuated does the pilot realize he is no longer flying level. Then, if the pilot instinctively pulls up, it makes the turn tighter and a stall can occur leading to an out of control spin, known as the "graveyard" spiral. Jensen flew to 4,000 feet, but his view was obscured by fog below and clouds above. There was no light above or below. After flying for a short while, he went into a tail spin, falling from the sky. His stunt flying experience paid off and he was able to bring the plane back to level. After a few minutes, he again went into a spiral downward. And, again, he fought to bring it out and succeeded just as he could make out the whitecaps on the water below. In all Jensen had three bouts of disorientation, three spins and three saves. He could have used Sloniger's homemade "hanging nut" to show whether he was flying level. Jensen lived through three close calls. He could not continue taking chances. Jensen decided that he must fly within sight of the ocean in order to stay level to the horizon. He set his altitude at one hundred feet utilizing his altimeter. Jensen periodically put his head out the window to see the whitecaps below.

Schluter was not able to calculate their route using the stars, so they continued using only the compass and calculated the route by "dead reckoning." Dead reckoning calculates one's position by taking into account the heading, speed, wind direction, and time since leaving a known fixed point. It is far less accurate than navigating by the stars. Jensen continued flying with his head out of the side window. His eyes strained to see the Pacific below. There was a thick layer of fog and clouds above. The night darkness was unrelenting.

Jensen said later, " I was flying by instruments. The altimeter said 100 feet, but at times the water was only 6 or 10 feet below." Jensen theorized that some anomaly caused the ocean to raise itself above normal sea level. At one point while flying low, Jensen inadvertently leaned against the control stick and *Aloha's* wheels hit the water. The water splashed up and put a rip in the fabric of the stabilizer. At that point Jensen raised his altitude to five hundred feet.

Jensen's fuel situation was dicey. *Aloha's* makeshift fuel tanks were filled with a hand pump taking fuel from free standing five-gallon containers to fill the main fuel tank. There was no fuel gauge. Each time Schluter transferred fuel to the main tank, he pumped until gasoline spilled out over the access cover. The spillage was then released into the ocean below. Valuable fuel was being wasted—to say nothing of the fire hazard. Jensen was a chain smoker.

Finally, almost two hours after Goebel had landed, Jensen brought *Aloha* in to Wheeler Field. He had five gallons of gasoline left, enough for twenty or thirty more minutes of flying time. He had averaged about eighty-six miles per hour as compared to Goebel's one hundred. Aloha's arrival was joyous, but subdued as compared to *Woolaroc's*. Marguerite Jensen who had been waiting along with Goebel collapsed with relief when she saw *Aloha* come in to land. After taxiing to a stop, Jensen remained seated for a minute. His first attempt to get up failed and he fell back into his seat. Finally, Jensen emerged from the plane looking haggard and red eyed. Schluter was seen slumped over in the back too exhausted to move. They had made it to Honolulu in just over twenty eight hours. Mrs. Jensen stood and asked, "Marty, where have you been?" She had recovered in time to join the celebration and the cheering by the crowd. Jensen remarked, "We got lost, but we made it. ...After four hours of wandering, we found ourselves and lit out like a blue streak for Wheeler Field."

It is somewhat surprising that Jensen made it. His factory fresh plane had not been in the air until six days before the race. Jensen's longest pre-race flight was about an hour. His gasoline situation was iffy. He literally flew by the seat of his pants. And yet, he succeeded when *Golden Eagle* and *Miss Doran* did not, even though those planes were better prepared for the race. Jensen demonstrated remarkable skill as a pilot.

On declaring Jensen the official second place finisher, James Dole remarked, "We're delighted that Jensen came in second. His appearance after he had believed by many to have been forced down, leads us to believe the others will arrive in good shape."

Richard DuRose

12

LOST AT SEA

After *Aloha* landed, the waiting crowd in Honolulu soon realized the other two contestants, *Miss Doran* and the *Golden Eagle,* might be lost. The crowd became nervous. The press pushed for information. William Malloska pointed out that Pedlar's plane had taken off last and had been airborne only twenty-two hours at that time, so it could be expected to arrive later. He said, "Boys, I'm telling you. Augie and Mildred will get here. Don't let anyone fool you on that score."

It had taken Jensen twenty-eight hours and sixteen minutes. *Golden Eagle* could fly no more than thirty hours. Auggie Pedlar and *Miss Doran* could remain aloft no more than thirty three hours. Those time limits came and went with no sign of either plane. The joy of seeing Goebel's and Jensen's success faded with the realization that Pedlar and Frost were certainly down somewhere in the Pacific Ocean. Neither carried a radio and neither had been seen by any of the ships along the route. Nor were there any safe alternative landing spots. Only the vast Pacific Ocean lies between Oakland and

Hawaii. The planes could be down anywhere along the route, or even off course if they had miscalculated.

James Dole, then George Hearst, sponsor of *Golden Eagle* and then William Malloska, owner of *Miss Doran* offered rewards to anyone finding the planes. A total of $40,000 was put up, with Dole offering $20,000 and Malloska and Hearst, $10,000 each. Dole was near collapse from despair.

The Federal Telegraph Co. broadcast to all ships. "Monoplane *Golden Eagle* and biplane *Miss Doran* probably out of fuel and landed in the ocean. Approximate location unknown. All vessels between the Pacific Coast of the United States and the Hawaiian Islands, especially those north or south of the regular lanes, are requested to keep a sharp lookout for floating planes, and if sighted, radio promptly full details address to the Federal Telegraph Co. of San Francisco, including exact location where found and condition of the occupants."

The awards dinner held at the Royal Hawaiian Hotel the night after Goebel and Jensen's landing was subdued. The guests wore black arm bands in respect for the missing fliers.

The Doran family gathered together awaiting news. As the hours passed, hope slowly slipped away. "We idolize

Mildred and we are so far away and so helpless to add our bit in the effort to find her," explained an aunt.

The Secretary of the Navy ordered a search from the Farallon Islands twenty seven miles off the coast of California to points fifty miles to the north and south of the Hawaiian Islands. Forty-two Navy vessels were deployed, including the aircraft carriers, *Langley* and *Arostook,* with several-dozen scout planes. Three of the Navy's submarines also joined the search. The Coast Guard deployed two ships and nineteen privately owned merchant vessels joined in. The Army Air Corps flew out fifty miles from Honolulu and San Francisco in the hunt. In addition, about one hundred Japanese fishing sampans joined the search in the area surrounding the Hawaiian Islands.

It was the most extensive Pacific search and rescue operation in history. It continued day and night for over a week. There was not a trace of either plane found. The Navy spent $125,000 in fuel and $40,000 for food for the eight thousand sailors involved in the search, close to $2,000,000 in today's dollars.

The Navy reportedly used its most up to date tactics in the search utilizing methods developed during World War I for sighting submarines. A Navy Commander described the operation: "Two men, standing in the crow's nest, one hundred and fifty feet above the water, scan the water on all sides. Two more lookouts are stationed in the wings of the bridge, together with an officer in each wing, and one on the bridge. Each lookout is assigned a quadrant. Working in two-hour shifts, because the strain on the eyes is...severe, the lookouts can see from twelve to thirteen miles in each direction, with the ship ploughing (sic) along at fifteen to twenty knots. Directed by Admiral Jackson the ships will be so deployed that the entire likely path of the flight, embracing more than 400,000 square miles, can be covered within three days." In total over 500,000 square miles were searched, an area roughly equivalent to the size of Texas and Oklahoma. Along the route many floating food crates and other debris discarded by the passenger and freight liners that sailed along the same corridor were found. But nothing was found of the fliers or their aircraft.

Newspapers covered the events of every day, repeating numerous rumors of sightings that turned out to be false. There was supposedly red white and blue wreckage floating near Maui. It turned out to be a colorful sampan. Flares of the type carried by the participants were spotted on Mona Kea on Maui. It was merely a farmer tending to his crops by moonlight.

Several newspapers proclaimed in giant headlines that the *Miss Doran* had been found with the occupants unharmed. William Doran, Mildred's brother, had been in the *Flint Journal's* newspaper offices when that report came in and hurriedly telephoned the news to the family. He dashed off a cable to his sister: "Dear Mil—I experienced the greatest thrill of my life when I heard you had been found. God Bless all of you. BROTHER BILL." The news report of a rescue was completely false, made up by reporters who failed to verify a rumor and hoping for a scoop. Another heartbreaking development occurred when the Doran family received a cruel message that for a fee, the unscrupulous caller would disclose her whereabouts. Gradually, the headlines echoed the disappointment and fear of the public that the planes and all those aboard were surely lost

The Flint Daily Journal reported the thoughts of William Malloska three days after the start of the Race. "Unshaven, eyes filled with tears, his trembling hands revealing his anxiety, William F. Malloska, backer to the plane, *Miss Doran* still clung to the hope that the plane and its crew might be found momentarily. In a husky voice, he said, 'I tell you. It makes a man feel queer, this thing does. Yesterday just knowing you would see the plane come in with the rest—and now—' He paused. There is just one thing that could have happened to prevent them from being picked up some place....that is, if the plane caught fire. 'That boat can't sink,' he declared. 'It can float forever. The plane was especially designed for such an emergency. It is a land plane, but built to float." He refused to discard the three leis he was saving for Miss Doran's crew.

Malloska had said to Mildred prior to the race, "If anything happens to you (and Augie), I will just wire for them to sell the business [Lincoln Petroleum Co.] to the highest bidder."

Some persisted in hope for the recovery of *Miss Doran* and the *Golden Eagle* Slonnie Sloniger pointed out that *Miss Doran's* fuel tank was rigged so that with the "sweep of a lever" all the fuel could be dumped in about eleven seconds, thus turning the tank into a flotation device. He declared, "The plane can float indefinitely." Herbert Hughes, manager at Buhl Aircraft echoed Sloniger's comments: "The buoyancy of the tank would support more than 2,700 pounds. The *Miss Doran* with its gas tank empty and without passengers weighs 2,100 pounds."

Golden Eagle was built with a water landing in mind. In addition to being the fleetest craft in the field, it could float. The tips of its wings contained rubber bladders holding air for floatation. The bottom of the fuselage was lined with balsa and cork. Just before the plane took off the pilot house and the navigator cabin and all other openings were lined with rubber to render them waterproof. A lever in the cockpit could release the landing gear to provide for a smoother landing on water. Its rubber life raft contained a sail. And, the plane's compass was removable so it could be transferred to the rubber raft.

After the planes took off on August 16, Bill Irwin continued to supervise repairs to the canvas covering of *Dallas Spirit*. The Race Committee proclaimed that unless a contestant took off by the 19th, it would not be an official Dole Race entrant. Livingston Irving removed his radio from the wreckage of *Pabco Pacific Flyer* and gave it to Irwin. On Friday morning, August 19, four-hundred-and-fifty gallons of fuel were pumped into *Dallas Spirit*. At 2:05 p.m., with about five thousand cheering onlookers, Bill Irwin took off for Hawaii. Irwin still sought the notoriety of being among the first to fly to Hawaii. As an additional justification for his late start, he vowed to search for the missing *Miss Doran* and *Golden Eagle* along the way. After consultation with the Navy, Irwin and Ivan "Ike" Eichwaldt pledged to take a zigzag course in order to enhance their chances of spotting the missing planes. Navigator Eichwaldt started sending messages by radio to

115

ships below giving a chatty account of the flight. Included in the messages were the following:

"2:20: Going strong. We are passing the docks and will see the lightship soon. We are carrying the tail high at 1700 feet and making close to 100 MPH speed. Will call again when passing the lightship.

2:25: We are passing Point Lobos. Now passing the lightship and see the two flag signals which means you bums are getting us. We can see the Farallones ahead.

2:50: We are flying at 300 feet and under the fog with a visibility of 30 miles and we are passing the Farrallones now.

3:33: Bill just had a drink of water.

4:20: We just passed close to a rain squall. The air was very bumpy. Visibility is clear ahead.

5:10: We just passed the SS Mana at 5:10 and dipped in salute. They answered us on their whistle. Of course we could not hear it but could see their steam. We might pick up to the squadron of destroyers before dark but that depends on the speed. All OK.

6:02: Tell the gentleman that fixed our lunch that it is fine except we cannot find the toothpicks.

7:12: We have 30 miles visibility and are flying at 900 feet. Have seen nothing.

8:00: It's beginning to get dark.

8:51: SOS ......Belay that we were in a spin but came out of it OK. We sure were scared. It sure was a close call. The lights on the instrument panel went out and it was so dark Bill could not see the wings.

9:02: We are in anr...."

The 9:02 broadcast was the last. Speculation is Eichwaldt was attempting to report another spin. The next day several ships, including the destroyer *Hazelwood,* searched the area where *Dallas Spirit* had broadcast the SOS, six-hundred-and-fifty miles west of San Francisco. Nothing was found. Erwin had pulled out of one spin, but most likely had not been able to do so the second time.

Around midnight on August 19, the freighter *West Sequana* spotted flares in the distance about two-hundred-and-fifty miles off the California coast. It wired the Federal Telegraph Company and headed toward the flares. By dawn they had found nothing and presumed it had been rum runners who had fled. *West Sequana* resumed its course to Shanghai in the morning.

On August 20, the Navy conducted an experiment by placing three men in a rubber raft similar to the one carried by *Miss Doran Golden Eagle* and *Dallas Spirit.* They discovered that although the seas were relatively calm, waves washed over the sides necessitating constant bailing. In addition, light winds blew the raft so that the ocean currents did not prevail, and its course was erratic. And, the Navy concluded the small craft was uncomfortably overcrowded with three passengers. As the hours and days passed the conclusion became clear; neither the crew of *Miss Doran* or *Golden Eagle* had survived.

The Navy first intended to search for a week. However, the fact that John Rodgers and his crew had drifted and sailed four-hundred-and-fifty miles in nine days, the Navy kept up a limited search for the downed planes for another two days before giving up On Thursday, August 25, all ships began to head for port. The last area to be searched was north and west of the Hawaiian Islands in case the fliers had overshot their target.

On August 20 and 21, organized prayers for Mildred Doran and the others were conducted throughout Flint. The *Flint Journal* reported, "A thousand factory men listened with bowed heads and grave faces this noon at Oak Park while prayer was offered for Miss Doran and the other missing Dole fliers. General prayers in factories, homes today and in the churches Sunday were requested by the Flint ministerial association."

On August 22, in a last desperate attempt to locate the *Miss Doran*, Malloska chartered a sixty-foot sampan to search the waters around Kuai, the area where Rodgers and his crew were found.

While there had been daily stories of the race beforehand, and in the first few days of the search, the press and the public began to lose interest. By that time newspapers had turned to the controversial news that Sacco and Vanzetti, proclaimed anarchists, had been executed on Tuesday, August 23, for murdering a paymaster in Boston.

Mildred's father, William A. Doran retreated to his farm with instructions not to contact him until the news was the "very best" or the "very worst." Her brother William E. Doran stayed at the offices of the *Flint Journal* pacing all night hoping to hear good news. He said, "Mildred will come through. She's too good to go that way." Little sister Helen, age twelve, continued to insist that Mildred would come home. She was the last family member to give up hope.

There was never one trace found of the three missing planes or their crews. There have been some reports that a piece of the wing of *Golden Eagle* was found. However, those reports only appeared in print years after the fact. Research failed to uncover any contemporaneous newspaper accounts of such a discovery. The Pacific Ocean devoured the seven aviators, as well as the aircraft carrying them.

Richard DuRose

13

CRITICISM

As the search wound down, there was second-guessing about the race, its preparations, and the fitness of the planes and their crews. In a published letter to the editor, Will Rogers, a well-known columnist and entertainer, (and aviation enthusiast) commented that all the planes should have been equipped with radios. Rogers later lost his life in 1935 in a plane crash. Ben Wyatt an inspector for the Department of Commerce said he had advised Pedlar not to take off after changing his spark plugs as he did not believe that mend was sufficient to correct the engine problems that had forced Pedlar to return. The Secretary of the Navy, Curtis Wilbur, proclaimed that in the future long distance flying "stunts" like the Dole Race should be more closely supervised and put under "rigid restrictions." On the other hand, Wyatt remarked that he was "at a loss to explain the non-arrival of Golden Eagle since it was the best equipped plane in the race."

Art Goebel and William Davis the winners of the Dole Derby wrote an article for the *Chicago Herald & Examiner*, summing up their complicated emotions: "What a cloud of despair enshrouds this little circle as we sit here beside our faithful old *Woolaroc* to write the first chapter of our story! Here we are, acclaimed as winners of the Dole race from America's mainland to Hawaii, and yet instead of feeling the exhilaration of that victory, we are only waiting impatiently, prayerfully, for the order that will permit us to go out on the ocean wastes to find our lost comrades. In the light of what has happened, we forget the hero-worship that has been forced upon us. We forget the material gain that has come from our quest. All that we can think of today is that we want to wind up our Whirlwind again and if we can but be granted another triumph from the fates, to soar out from the scenes of our own little happiness and give all—if necessary—that our less fortunate friends may be brought safely into port."

It seems likely that the *Miss Doran* and the other planes went down in the Pacific Ocean due to any of four reasons: 1) a shortage of fuel; 2) going off course until all fuel was spent; 3) falling into a graveyard spiral; or 4) engine failure. There is evidence that any of them could have been the fatal flaw. As to fuel, Augie Pedlar was observed dumping fuel on his way back to Oakland to receive repairs. The fuel could be entirely jettisoned from *Miss Doran* in only eleven seconds. He could have forgotten to replenish the tanks prior to the second takeoff. As evidenced by the Ernie Smith flight, fuel calculations were imprecise and there seemed to have been a tendency to underestimate fuel consumption.   Second, the compass on the *Miss Doran* misbehaved in the days prior to the race. The plane had no radio. Miss Doran may have flown off course and missed Hawaii altogether. Third, the planes in the race were susceptible to spinning out. Since Jensen and Irwin both reported foggy and cloudy conditions, which caused them to spin, there is good reason to surmise that Pedlar (and perhaps the others) would have experienced the same issue. Indeed, Slonnie Sloniger speculated that this could have been a problem with Pedlar as he was relatively inexperienced. Finally, just days before the race, Augie Pedlar, on the relatively short hop from San Diego to Oakland, made an emergency landing in a wheat field due to a sputtering engine. He changed the spark plugs, but had similar engine problems a week later after his first Dole Race takeoff. When he returned to Oakland, the mechanics urged Pedlar to withdraw from the race. They questioned whether another change of spark plugs was sufficient to rectify the extreme misfiring of the engine. Based on this evidence, it could be that during the flight, the engine just gave out.

After eighty plus years, the mystery of why the planes went down will never be known. In the end the exact cause is not important. Augie Pedlar, Cy Knope, and Mildred Doran along with four crewmembers from the other planes, perished during their quest to cross the Pacific from California to Hawaii because the planes and the aviators were not sufficiently prepared for the dangers of such a long flight.

In all the Dole Race cost the lives of ten aviators. Three died in crashes before the race. Five died during the race. And, two more died in a flight that began three days after the race. The Navy and the Army spent hundreds of thousands of dollars in fuel and equipment during their exhaustive search. Many people questioned whether the race had been worth it.

It was the beginning of the air age, and the Dole flyers let their enthusiasm for fame and adventure get the best of them. Lindbergh had lit a fire. The public had become "air minded." Most of the participants seemed to be counting on good luck rather than careful planning. Much was unknown about attempting such a long 2,400 mile flight over water, including how to measure fuel consumption, navigating in foggy, and cloudy conditions, and the danger of disorientation when "flying blind." The eleven weeks given the racers to prepare was too short. The fliers were long on ambition. They did not have experience to realize what they lacked in wisdom. And to their credit, Goebel and Jensen proved it could be done, although one suspects even they realized they were very fortunate to have made it.

Richard DuRose

14

Tributes And Remembrances

On September 14, a memorial service attended by over a thousand was held at Pier 30 in San Francisco in honor of the seven lost Dole aviators. Cy Knope 's wife, Gwendolyn, and young two-year old daughter attended. Prayers and soothing words were offered by a Catholic priest, a Navy chaplain, a Protestant pastor and a Jewish rabbi. A large choir sang "Nearer My God To Thee." Many of those attending donated bouquets or wreaths. There were five thousand floral offerings in all. At noon, as a lone bugler played *Taps*, the steamer *Maui* slowly crept away from the dock for Honolulu on its regular voyage to the Islands. As the steamer slipped away, Hawaiian singers with ukuleles sang *Aloha Oe; Farewell to Thee*.

After two days and seven-hundred-and-fifty miles, the ship's engines were shut down and the passengers and crew assembled on the forward deck for another formal memorial service. The twenty third psalm, "The Lord is my Shepherd..." was read. The hymn, " Going home, going home, I'm going home..." to the melody of the Largo movement from Dvorak's *New World Symphony*, was sung Walter Cribbens gave an emotional eulogy praising the seven lost souls. He remarked, "I believe that the beauty and sweetness of Mildred Doran have been preserved as ornament and perfume [in] that house of the many mansions. I believe that the smiles of affection and deeds of kindness, and the pleasant memory which she has left with us will be cherished forever." Also giving a tribute was Rabbi Louis Newman. Flowers were flung into the water. One garland, in the shape of an open Bible, was contributed by school children from Caro, Michigan, where Mildred Doran taught. It bore an inscription, "God Bless You, Every One," Tiny Tim's wish from *The Christmas Carol.* The array was designed with a cork base to allow it to stay afloat indefinitely.

The Boston *Daily Advertiser* published an editorial that was re-printed in other newspapers around the country. Under the caption, "Mildred Doran, womanhood's first martyr to the air conquest of the great oceans," it commented, "In all likelihood, Miss Doran, the flying schoolma'am is dead. But while she lived, she lived greatly. That is more important than length of days... When the spark has passed from the bodies of all who read these lines, admirers of valor and of noble spirit will be celebrating the name of Mildred Doran. This pretty, fearless girl, a slip of a lass with a heart courageous will live in the minds of men and women when most of us have been forgotten...."

On October 4, 1927, the New York *Times* reported in a short article that a note in a bottle had been recovered "near Albany, opposite the Golden Gate... on the San Francisco Bay." The note, which was purportedly written by Mildred Doran, indicated that the engine is "still missing" and that the plane had been in a "nose dive." But, as the article points out, the note contained several obvious mistakes in grammar — definitely not something a school teacher would have written.

Of course, there still are unexplained and tragic crashes of airplanes. In 2008, the National Transportation Safety Board (NTSB) reported one-hundred-and-thirty-six fatalities from crashes of small private planes. Worldwide there are an average of five-hundred-and-seventy-three deaths annually from scheduled airline crashes. There have been innumerable safety innovations since 1927. Airplane design is vastly improved. But gravity still takes its toll.

Air shows continue to be a summer attraction throughout the United States and the world. Unfortunately, stunt pilots and wing walkers continue to lose their lives by taking chances just as they did ninety years ago. Over 70,000 spectators saw Todd Green fall to his death attempting to climb from the fuselage of a small plane onto the rail of a helicopter at the Selfridge Air Show in August 2011. The stunt and the resulting misfortune are reminiscent of The Twenties.

Regular service by air between the west coast and Hawaii began in 1946 after World War II. As fate would have it, the pilot of the first passenger flight was "Slonnie" Sloniger, the pilot who lost the toss of a coin in 1927. Matson Steamship Co. had given up its ships to the war effort. Without any ships, it had to find other ways to generate income. It went into air service. The DC-4 carried a crew of four in addition to the pilot, including a co-pilot, a radio operator, navigator, and flight engineer, as well as several stewardesses. Matson chose a military-like uniform with four stripes on the sleeve, like a ship's captain, for its pilot, a strategy later copied by other airlines.

In Flint, Michigan, flags were flown at half-mast. On Thursday evening, August 16, a memorial service at Mildred Doran's church, Lakeview Methodist Episcopal Church on East 15th Street was conducted. Mildred's two brothers, William and Floyd, and her younger sister, Helen, described in news accounts as a "second Mildred," attended. Reverend R. H. Prouse presented the eulogy and noted that heroism is defined by "...the courage, the vision to go on when others have quit the race; to get a little further than other people are willing to go. ....Women of the world ought to be proud of one who had daring and courage to blaze a trail."

In January 1928, the Ontario (Canada) Department of the Interior named some lakes in northwestern Ontario for aviators who had lost their lives in attempting transoceanic flights. One large lake in the territory north of the Red Lake District was named "Doran Lake" in honor of Mildred Doran. According to the Canadian Natural Resources Department, the name was proposed by F.H. Peters, Surveyor General. The lake is situated south of Lake Joseph at 50:58:22 N Latitude and 90:34:50 W Longitude.

The Anchor, a publication of Alpha Sigma Tau sorority, published a tribute to Mildred in March 1928, in which she is described as a "quiet, modest young girl, friendly, considerate of all those about her, and enthusiastic over the great venture about to be taken."

Addison N. Clark, a mining engineer from San Francisco sent the Doran family an original poem, "Into the West." Its final stanzas express a melancholy thought:

> *Four fleet falcons with man-made power,*
> *Thundered, a hundred miles an hour,*
> *Bearing eight men and a woman fair,*
> *On through the miles of trackless air*

Richard DuRose

Into the West

All through the night of hopes and fears,
Through racking hours that seem like years,
Kinfolk, lovers, old friends and new,
Waited for word as on they flew
Into the West

"Goebel wins!" barks the radio wave,
"Jensen second!" the newsboys rave,
Theirs the joy of the earned award,
But — what of the other souls that soared
Into the West"

Pedlar and Knope and Scott and Frost,
Intrepid Mildred, we know you're lost.
To us who live on dusty ground,
But, perhaps we envy the path you found
Into the West.

In October 1927, the popular country singer, Vernon Dalhart recorded a song about Mildred. Dalhart is credited with the first million selling record, "The Prisoner's Song." *If I had the wings of an angel, over these prison walls I would fly..."* Dalhart recorded *The Fate of Mildred Doran* under his own name and, as was his custom, using another name, Al Craver. Among its lines,

"This is the song of a maiden fair.

No man could be braver than she.
 For she gave her life for the country's cause,
When her airship went down in the sea...

 Sad were the hearts of the world that night,
When they heard of the fair maiden's fate.
 And her friends once again seemed to see her smile,
As she flew thru the old Golden Gate...

 For life is uncertain with all,
And we should be ready to meet our fate.
 For we know not when death may befall."

In 1929, William Malloska, constructed a building at the intersection of South Saginaw and Maple Streets in Flint, on a corner of the Lincoln Airfield. It was a six-sided, three-story building, tapered toward the top resembling a wind mill structure without the sails. It housed a gas station on the first floor and an apartment on the second. In an elaborate ceremony it was formally named the "Doran Tower." On a large boulder next to the building was a brass plate inscribed to read, "Only those are fit to live that are not afraid to die," a quote from Theodore Roosevelt, later repeated by Douglas MacArthur. At the time of its dedication, Malloska proclaimed that the memorial would stand for one hundred years.

In 1930, Lincoln Oil was sold to Cities Service Oil Co. (now Citgo). Subsequently, road improvements at the Doran Tower location resulted in the plaque being lost. In 1932, the land on which the airfield stood was sold to a real estate housing developer and it became the Lincoln Manner Subdivision. As a result, the Doran Tower was relocated. In later years, the gas station became a restaurant and then a flower shop. In 1972, the owners were required to pay a sewer tap-in fee that they could not afford. In 1973, the Doran Tower was torn down. The tower stood only forty three years. Malloska died in obscurity in Brownsville, Texas in 1941.

Among the remembrances sent to the Doran family is a poem from Marcel Stani Ducourt, a writer for *L'Aerophile,* the journal of the Aero Club de France in Paris the world's oldest association of aviators. The poem is written in French with an English translation by Ducourt. It begins:

> *"She had grey eyes sharpened,*
> *Toward the temples,*
> *With lips from red,*
> *Was often bitten by the teeth.*
> *She had a thin and easy feline body.*
> *An American girl, simple and complex,*
> *Whose light soul vibrated,*
> *With desperate desires..."*

And it ends as follows:

> *"That's why she had gone,*
> *In the womanly sweetness,*
> *Of a Californian night,*
> *On this Pacific Ocean.*
> *She never went back.*
> *Her name was Mildred Doran."*

Another of the poems received by the family is by Doris Skewes titled, The *Crew of the Miss Doran*. Its first verse reads as follows:

> *Behind them vanished shorelines gray,*
> *Behind them stretched the Golden West.*
> *Before them glowing conquest lay,*
> *Before them a game their souls to test.*
> *The lone girl said: "Now must we pray,*
> *For lo! The very stars are gone.*
> *Brave Pilot, speak! And let us say,*
> *'O God! May we fly on and on!'*

Ms. Skewes indicated that her poem was based on the famous poem *Columbus*, written by Joaquin Miller. *Columbus* was widely memorized and recited by school children.

One cannot help but wonder if Mildred recognized the dangers of a 2,400 mile flight in an airplane over the Pacific

Ocean in 1927. Until then, Rodgers, Maitland, and Smith had all survived, even though they had difficulties. So it was generally recognized that there were dangers. Then, in the days leading up to the race, three planes crashed, killing three aviators. All of the Dole fliers must have pondered those tragedies.

But what was Mildred thinking at the time? We shall never know her innermost thoughts. Had Mildred fallen in love with being a celebrity and lost her perspective? Was she overtaken by a sense of obligation to carry through? She had a dedication to duty. She had taken care of her siblings. She seemed to be self-assured, but humble in her comments to the press, and she spoke of the danger in broad terms. Her father had proclaimed that Doran's were not yellow. From June to August, she told the whole nation that she was going to attempt the flight. Then, the *Miss Doran* was forced to return to Oakland with a sputtering engine. What a danger sign! It is difficult to draw a line between courage and recklessness. There must have been doubts prior to the second takeoff, especially because Pedlar and Knope urged her not to go. In retrospect, one cannot help but wish she *had* backed out. It was just the beginning of the story of women aviators. In 1929, just two years later, The Ninety Nines, Inc., an international association of female pilots was formed by ninety-nine charter members, including Amelia Earhart. That same year twenty women participated in The Women's Air Derby, a race from Santa Monica to Cleveland. While it is true that Mildred probably inspired some of the female pilots that followed her, one cannot help but think that Mildred would have reveled in the deluge of women's flying activities just ahead. It would have been far better for Mildred to lose face than to lose her life. And, I really would have liked to have known her as Aunt Mildred.

During the build up to the Dole Race, Mildred was as visible as the unexpected bright flash of a meteor racing across the night sky. But, also like a shooting star, she abruptly disappeared from view forever. No matter how spectacular, we do not remember a "Shooting Star" for long. Nor, has the public remembered Mildred Doran. I hope this book will make a difference.

Richard DuRose

EPILOGUE

My mother was Mildred's young sister, Helen Doran. She was only twelve when the Dole Race was held. The two sisters had lost their mother to tuberculosis eight years previously. Within a year of the Dole Race, their father died as well. I was born ten years after the Dole Race. My mother never spoke of Mildred except to generally acknowledge that she had been lost at sea in an airplane race. I was ten or twelve when I first saw the Doran tower. I do not recall that my mother ever visited that monument. It is obvious from Mildred's comments in news stories that she had a strong affection for her younger sister. I believe now that my mother chose not to think about or re-live the heartbreaking loss of her sister.

As an aside, my mother pronounced her maiden name, "Dor'-an" with an emphasis on the first syllable, not the second as I have heard some pronounce it.

The Dole Race was never talked about as a part of our family history. As a result, I never had much curiosity about it. My parents died over twenty years ago and now there is no one to ask. For reasons not clear to me, my mother never spoke to her other brother, William E. Doran, a probate judge in Flint at any time during my lifetime. Apparently it was a family squabble that occurred before I was even born. As a result, I never did learn of the events of 1927 through the family.

Curiously, 1951 newspaper accounts relate that Judge Doran was disbarred for misusing $5,184 from the estate of William Malloska who had died ten years earlier. Doran admitted taking the funds because he was in "financial straits." He reimbursed the money and was later reinstated to the bar. In preparation for this book, I went through the pictures, albums, and memorabilia my mother had collected. I had never looked at them carefully. I noticed that in addition to pictures of Mildred, there are photos of my mother at age twelve with William Malloska Slonnie Sloniger, and another Lincoln Standard pilot, Bill Atwell. She must have caught the flying bug as well.

Included in my mother's papers were various pictures of Mildred, some in everyday poses with cars or horses, and some from the Dole Race in which she was dressed for flight. As reported above, her dog, Honolulu, had to be left in Oakland. There is a picture in the album of the dog taken five years later. There was the postcard from St. Louis and the letter from San Diego that are quoted above. Also, in the packet were about three hundred newspaper clippings from Flint and Detroit, as well as from across the U.S., regarding Mildred and the Dole race. Unfortunately, many of them were clipped in such a way that the identity of the paper and the exact date of the article is not shown.

Once I started to look into the story, I could not stop. The story of the Dole Race is fascinating. So what you have in this piece is what I learned. My mother's older sister, Mildred, was pretty, bright, outgoing, and what we would call a media star today. News reports quoted her at each stop from Flint to Oakland on a daily basis. The entire country followed her quest. Then in a flash she was gone; disappeared. While my mother was a normal happy person, she hardly ever spoke of her sister. I am sure the pain of losing her sister was deep. And, curiously, my mother never learned to drive a car because, she said, it made her too nervous.

Richard DuRose

The Dole Racers

1. OKLAHOMA – A Beech Travel Air cabin monoplane with a blue fuselage and orange wings. Wing area: 312 sq. feet. Fuel capacity: 425 gallons. Pilot: Bennett "Benny" Griffin. Navigator: Al Henley. Took off on August 16, 1927, at noon, but returned shortly after passing the Golden Gate Bridge with five blown cylinders and thick black smoke streaming from the engine. The blown engine was too extensive to repair.

2. EL ENCANTO – A silver Goddard design. Wing area: 283.5 sq. feet. Fuel capacity: 360 gallons. Pilot: Norman Goddard. Navigator: Lieut. K.C. Hawkins. Attempted takeoff at 12:02 pm. on August 16, 1927, but crashed when it ground looped, severely crumpled a wing, and was unable to continue. Goddard and Hawkins survived unhurt.

3. PABCO PACIFIC FLYER – An orange and black Breese monoplane. The word "Pabco" is a trade name used by its sponsor, The Paraffine Company. It also sported the Indian Head insignia of the Lafayette Escadrille for whom Livingston Irving served in WWI. Wing area: 260 sq. feet. Fuel capacity: 380 gallons. Pilot: Major Livingston "Jimmie" Irving. Navigator: none. Irving aborted his attempt at during his first takeoff at 12:05 pm. on August 16, 1927 at 12:33 pm. On the second takeoff a few minutes later, it ascended to 20 feet, came down, ground looped, and crashed. At this point, Irving gave up. Irving gave his radio to *Dallas Spirit's* Bill Erwin for his rescue attempt two days later.

4. GOLDEN EAGLE – A Lockheed Vega cantilever (no wing struts) monoplane painted gold, with a blue landing gear. Wing area: 260 sq. feet. Fuel: 350 gallons. Pilot: John "Jack" Frost. Navigator: Gordon Scott. Took off on August 16, 1927, at 12:31 pm. and was lost over the Pacific without a trace.

5. *MISS DORAN* – A Buhl Air Sedan biplane with a red nose, white fuselage, and blue tail. Wing area: 350 sq. feet. Fuel Capacity: 400 gallons. Speed: 120 mph. Pilot: John A. "Augie" Pedlar Navigator: Cyrus "Cy" Knope. Passenger: Mildred Doran. Took off from Oakland on August 16, 1927, at 12:32 only to return with a sputtering engine for repairs. After almost two hours of mechanical work, she took off again at 2:03 pm. The crew and plane were lost over the Pacific without a trace.

6. *ALOHA* – A bright yellow Breese monoplane with a blue and red depiction of a lei around her nose and the word "Aloha written in oversized letters on both sides. Wing area: 260 sq. feet. Fuel: 390 gallons. Pilot: Martin "Marty" Jensen. Navigator: Paul Schluter. Took off August 16, 1927, at 12:34 p.m. and finished second behind Goebel by about two hours with only 5 gallons of gas remaining.

7. *WOOLAROC* – A Beech Travel Air cabin monoplane with a blue fuselage and yellow wings. Wing area: 310 sq. feet. Fuel: 395 gallons. Pilot: Art Goebel, Navigator: Lieutenant Bill Davis. Took off at 12:35 p.m. on August 16, 1927 and landed at Wheeler Field Honolulu after a little more than twenty-six hours. It was the winner of the Dole Derby.

8. *DALLAS SPIRIT* – A Swallow with a dark green fuselage and silver wings. Manufactured in Wichita, Kansas. Wing area: 330 sq. feet. Fuel capacity: 480 gallons. Pilot: Bill "Lonestar Bill" Erwin. Navigator: Ivan "Ike" Eichwaldt. Took off at 12:37 p.m. on August 16, 1927, but returned with a large rip to the canvas covering its fuselage. Erwin took off again on August 19, 1927, vowing to search for missing planes. Livingston Irving donated his radio to Erwin for the flight. During its flight, Eichwaldt broadcast several radio messages, but after 650 miles there was silence. No trace of the plane and crew was found.

9. CITY OF PEORIA – An Air King. Wing area: 342 sq. feet. Fuel capacity: 372 gallons. Pilot: Charles W. Parkhurst. Navigator: Kenneth Lowes. This blue and silver open cockpit biplane was owned by the National Airway System, which had begun manufacture of the Air King in Lomax, Illinois in 1926. E. M. "Tex" Lagone served as the owner's representative and accompanied the plane to Oakland. It was disqualified for too little fuel capacity on the morning of August 16, 1927, shortly before the noon start of the Dole Race. The plane was scrapped shortly after the Dole Race.

10. HUMMING BIRD a/k/a The Spirit of John Rodgers—A Tremaine-Thalfield Pacific J-30 low winged, monoplane. Engines: British-Lucifer. Only plane in the field without Wright J-5 engines. Pilot: George Covell. Navigator: Richard Weggener. Crashed in a fog into an embankment at Point Loma, California, on its way to Oakland from San Diego's North Island Navy Base on August 10, 1927. The resulting fire killed both Covell and Weggener.

11. SPIRIT OF LOS ANGELES – A Fisk International triplane with two engines. Pilot: James Giffin. Some news reports incorrectly spell his name Griffin. Crashed into the bay at Oakland on August 11, 1927, five days prior to the race during a practice run when it missed the runway. The pilot and two others aboard were fished out of the Bay, but not hurt. The plane was too damaged to continue.

12. ANGEL OF LOS ANGELES – A Bryan cantilever monoplane with two engines. Pilot: Arthur V. Rogers. Navigator: Leland Bryant. Crashed during testing in Montebelo, California, on August 13, 1927. Rogers, the only occupant at the time, jumped out, but his parachute did not open. He was killed after a fall of 150 feet to the runway to the horror of his wife who was observing the flight.

13. MISS HOLLYDALE — Designed by Edwin Fisk. Biplane, manufactured by International Air Craft of Long Beach, California. Pilot: Frank Clark (sometimes erroneously spelled Clarke). Navigator: Jeff Warren. This black and orange biplane was made of plywood and had a 35-foot wingspan. Frank Clark disagreed with the decision to postpone the August 12 start date, and on August 13, took off from Oakland without any explanation, fueling speculation that he was heading to Honolulu. The next day, his benefactor wired the Race Committee that Clark had flown back home to Los Angeles and was withdrawing from the race.

14. DETROIT'S GOODWILL MESSENGER — AHess Bluebird. Pilot: Frederick Alexander Giles, an Australian airmail pilot. Owner: W. H. Rosewarne, a Detroit contractor. Giles was unable to get the plane to Oakland in time for the race.

REFERENCES

The Anchor, Alpha Sigma Tau Sorority, March 1928: "Mildred Doran, a Tribute."

Associated Press, August 9, 1927: "Girl Flyer's Laundry Waits In Honolulu."

Associated Press, August 11, 1927: "Pedlar May Hop Off Before Other Pilots."

Associated Press, August 13, 1927: "Arthur Rodgers (sic) Crashes In Test At Los Angeles."

The Bulletin, San Francisco: Race Souvenir Edition, undated.

Honolulu Star Bulletin, Burlingame, Burl, December 29, 2003, January 5 & 12, 2004: "The Dole Race."

Chicago Herald & Examiner, August 19, 1927: "Dole Race Winners Eager To Aid Hunt."

Chicago Herald & Examiner, Extra, August 20, 1927: "Miss Doran And Her Pilots Found Safe in Hawaiian Bay."

The Chronicle, San Francisco, California, September 15, 1927: "Maui Bears 5,000 Floral Pieces For Mid-Sea Tribute."

The Citizen, Ottawa, Canada, August 19, 1927: "Eganville, Ontario Born Girl Reported Missing."

The Nevada Daily Mail (Nevada, Missouri), November 23, 1927: "Giles Returns To San Francisco."

The Day, (New London, Connecticut), August 16, 1927.

Davis-Monthan Aviation Field Register, Tucson, Arizona, July 19, 1927.

Detroit Free Press, August 18, 1927: "Despair Of Doran Kin Lightened By Rumors."

Detroit Times, July 18, 1927, Doran, Mildred A.: "I'll Make Hawaii: Miss Doran Says."

First Stop Honolulu, Ted Scott Flying Stories, Dixon, Franklin W.: Grosett & Dunlap, 1927.

San Francisco Call Bulletin, October 10, 1955, Eshleman-Conant, Jane: "Pioneer Pacific Fliers Wrote Tragic Chapter In Air History."

The Evening Independent (St. Petersburg, Florida), June 8, 1927: "With About 30 Flyers Unofficially Entered, San Francisco Prepares For Honolulu Flight."

The Flint Daily Journal, August 8, 1927: "Flying Teacher Wins 4th Position in Hawaii Race."

The Flint Daily Journal, August 19, 1927: "Malloska's Belief in Plane Unshaken."

Flint Journal, September 2, 2008, Flynn, Gary: "Memorial Lost, Just Like Air Pioneer Mildred Doran."

Glory Gamblers, Forden, Leslie, Ballantine Books, 1961.

Art Goebel's Own Story, Goebel, Art (Fourth Edition), 1929; edited by G.W. Hyatt.

Aviation History, September, 2005, Grover, David H.: "Emory Bronte and Ernie Smith: Flew From California to Hawaii in 1927."

Hawaii Aviation, An Archive of Historic Photos and Facts, Chapter VI, All The Way; Chapter VII, Enter the Civilians.

Above The Pacific, Horvat, William J., Aero Publishers, 1966.

International News Service, August 18, 1927: "Fliers Confident Parting Words."

California State Library Association, Bulletin No. 93, Kurutz, Gary: "Triumph and Tragedy Over The Pacific: The 1927 Dole Race."

New York Times, August 16, 1927: "Crowd Sees Crashes At Start," "Four Planes Winging Across Pacific," and "Ships Report Planes At Sea."

New York Times, August 19, 1927: "Navy Leading Great Hunt," "Kin Grow More Anxious," and "Mother Blames Engine."

New York Times, August 20, 1927: "Ideal Weather Aids Hunt," and "Flint Citizens Pray For Their Girl Flyer."

New York Times, August 21, 1927: "Navy Test Of Raft Lessens Hopes For Survival In Pacific Of Missing Fliers."

New York Times, October 8, 1927: "Note From Miss Doran Is Found In Bottle."

New York Times, November 24, 1927: "Doubt Storm Ended Giles' Hawaii Hop, San Francisco Weather Experts Say Conditions Flier Reported Were "Impossible."

New York Times, January 14, 1928: "Ontario Names Lake For Missing Fliers."

The Record Eagle, Traverse City, Michigan, December 5, 1961: "Probate Judge Found Guilty."

Robison, Carson, "The Fate of Mildred Doran" a/k/a "Mildred Doran's Last Flight" recorded by Vernon Dalhart and distributed on Columbia Records, and on Challenge Records by Al Craver, one of Dalhart's many aliases.

The San Diego Union, August 3, 1927: "Five San Diegans Enter Hawaii Race."

Pacific Air Race, The true story of the first Pacific air race, August 16, 1927, *Smithsonian* Institution Press, 1988, Scheppler, Robert H.

One Pilot's Log, The Career of E.L. "Slonnie" Sloniger Sloniger, Jerrold E., Howell Press, 1977.

United Press, August 22, 1927: "Tragic Air Derby Sponsor Near Collapse."

The World, (New York) August 18, 1927: "Honolulu Gone Race Wild."

Associated Press, August 22, 2011, Willimas, Corey. "US Probes Wing Walker Death At Michigan Air Show.

Acknowledgements

While I spent my career as a lawyer writing for courts and other lawyers, writing this book has been a brand new experience. I want to thank Patricia Trenner, Senior Editor at Smithsonian's Air & Space magazine who accepted a short version of Mildred Doran's story for that publication, and who encouraged me to continue with this book. Also, I cannot express adequately my gratitude to my dear friend, Mary Kauffold who spent many hours consulting with me on the organization and content of the book. Finally, I must express my thanks to Nancy DuRose, my wife who answered the question, "How does this sound?" at least a thousand times. Also, thanks to the many acquaintances that read the Air & Space article and said, in effect, "You should write a book."

Richard DuRose

About The Author

Richard DuRose, formerly a corporate lawyer specializing in labor and employment in Florida and Ohio, lives in Hendersonville, North Carolina. He enjoys hiking and golf.

He has been researching the story of the Dole Race and his Aunt Mildred for over three years and continues to be interested in learning the stories of the participants of that race.

He may be contacted through his Email at *rdurose@morrisbb.net.*

Richard DuRose

3 1901 05413 0895

Made in the USA
Charleston, SC
06 September 2011